Allied Participation in Operation IRAQI FREEDOM

by

Stephen A. Carney

Center of Military History
United States Army
Washington, D.C., 2011

Published by Books Express Publishing
Copyright © Books Express, 2012
ISBN 978-1-78266-107-8

Books Express publications are available from all good retail and online booksellers. For
publishing proposals and direct ordering please contact us at: info@books-express.com

Contents

	Page
Introduction	vii
Allied Participation in Operation Iraqi Freedom	1
Overview	1
Formation of the Coalition of the Willing, November 2002–March 2003	5
Major Combat Operations: Coalition Forces Land Component Command, March–May 2003	6
Combined Joint Task Force–7, June 2003–May 2004	12
Multi-National Force–Iraq, May 2004–July 2009	18
Analysis	30
Appendix—Force Contributions by Nations	34
Albania	35
Armenia	37
Australia	39
Azerbaijan	42
Bosnia-Herzegovina	45
Bulgaria	47
Czech Republic	49
Denmark	52
Dominican Republic	55
El Salvador	57
Estonia	60
Georgia	62
Honduras	65
Hungary	66
Italy	68
Japan	71
Kazakhstan	74
Latvia	76
Lithuania	78
Macedonia	80
Moldova	82
Mongolia	84
The Netherlands	86
New Zealand	90

Nicaragua 93
Norway 95
Philippines 97
Poland 98
Portugal 101
Republic of Korea 102
Romania 105
Slovakia 108
Spain 110
Thailand 113
Tonga 115
Ukraine 116
United Kingdom 119
Further Readings 125
Abbreviations 127
Map Symbols 129

Tables

No.
1. Coalition Partners in Iraq, 2003–2009 . . . . *f.* 32
2. U.K. Unit Rotations, 2003–2007 120

Charts

1. Coalition Forces Land Component Command and U.S.
   Combined Forces Special Operations Component
   Command as of May 2003 8
2. Combined Joint Task Force–7 as of June 2003 15
3. Multi-National Force–Iraq as of June 2004 20
4. Multi-National Force–Iraq as of June 2005 23
5. Multi-National Force–Iraq as of June 2007 25
6. Multi-National Force–Iraq as of June 2009 30

Maps

1. Operation Iraqi Freedom, 20–28 March 2003 10
2. Operation Iraqi Freedom, 29 March–1 May 2003 11
3. Operation Iraqi Freedom, Combined Joint Task Force–7,
   June 2003–May 2004 16
4. Operation Iraqi Freedom, Multi-National Force–Iraq,
   June 2004–May 2005 21

No. *Page*

5. Operation IRAQI FREEDOM, Multi-National Force–Iraq,
   August 2008 28
6. Operation IRAQI FREEDOM, Multi-National Force–Iraq,
   July 2009 31

Illustrations

General Tommy R. Franks 7
Lt. Gen. Ricardo Sanchez and Secretary of Defense Donald H.
   Rumsfeld 12
Retired Lt. Gen. Jay Garner and Secretary Rumsfeld 14
General George W. Casey Jr 18
General David H. Petraeus 26
Lt. Gen. Raymond T. Odierno 29
Soldiers from an Albanian Komando regiment provide security
   along a roadside in Mosul 36
Coalition forces commemorate the end of mission for the
   Armenian contingent 38
Australian Army soldiers on patrol in Iraq 40
An Azerbaijani soldier closes the gate at entry control
   point 1, Hadithah dam 43
Explosive ordnance disposal soldiers of the Army Forces of
   Bosnia-Herzegovina prepare the day's find for
   detonation after a sweep near Tallil 46
Czech Republic soldiers explain to an Iraqi Army tank crew
   member how to zero the T72 tank at the Besmaya
   Gunnery Range 50
During a training exercise on the outskirts of Al Basrah, a
   Danish Army soldier inspects a target vehicle as
   other troops prepare to fire their weapons 53
Soldiers of El Salvador's Cuscatlán Battalion wait to distribute
   food at an Iraqi displaced persons camp near Al Kut .. 58
A soldier from the 13th Georgian Army Light Infantry Battalion
   provides security from behind a wall during a patrol
   through Al Lej 63
Hungarian troops greet Iraqi civilians 67
Italian and Iraqi soldiers provide security at a meeting between
   Italian, Iraqi, and U.S. military leaders in An
   Nasiriyah 69
A Japan Air Self-Defense Force soldier salutes before opening
   the gate to a flight line 72

Iraqi soldiers pass explosives down to the site of a controlled
    detonation by Kazakh soldiers and Iraqi explosive
    ordnance disposal trainees on Forward Operating
    Base Delta 75
U.S. Army and Latvian soldiers provide security for a convoy near
    Tallil 77
A Lithuanian soldier oversees activity in a market in the town of
    Ad Dujayl 79
A soldier from the Netherlands provides force security at Tallil
    Air Base 88
New Zealand engineers busy at work on the At Tannumah
    Bridge, Al Basrah 91
Norwegian Army troops on patrol in Iraq 96
A Polish Army unit conducts a convoy briefing before heading
    out on a mission from Camp Echo. 99
An Iraqi civilian, selected to receive free medical care, is cheered
    on by soldiers from the Republic of Korea at Camp
    Zaytun, Arbil. 103
The color guard for Romania's 26th Infantry Battalion marches
    during the end-of-mission ceremony at Camp Dracula. . . 106
Spanish Army soldiers are among the multinational coalition
    forces watching a ceremony at Camp Babylon. 111
Soldiers from a Ukrainian Army nuclear, biological, and chemical
    unit discuss their techniques and equipment with coalition
    partners at Camp Arifjan, Kuwait 117
A British investigator with a joint U.S.-British bomb examination
    team studies the wreckage from a car bombing near Camp
    Rashid 121

All illustrations are from the files of the Department of Defense.

Introduction

The invasion of Iraq in March 2003—Operation IRAQI FREEDOM—was controversial at its start. The United Nations was reluctant to provide a specific endorsement for direct U.S. military action. Without this authorization, a number of close allies refused to participate in the operation. In order to garner greater support and provide an international flavor to the intervention, President George W. Bush assembled a "coalition of the willing," ultimately involving about sixty nations. Although some of these countries supplied little more than nominal assistance, fully thirty-seven of them furnished a total of around 150,000 ground forces from the start of the operation through July 2009. These troops conducted security operations; provided reconstruction assistance; operated command-and-control headquarters; and fought, were wounded, and killed alongside U.S. soldiers, sailors, airmen, and marines. This temporary alliance was more than just a paper coalition; it involved substantial and important support from our international partners in helping achieve U.S. war aims. It is important that the United States Army and the American people know about and remember the sacrifices of these allies.

Allied Participation in Operation IRAQI FREEDOM highlights a number of key aspects of allied support to the U.S.-led operation. The presence of ground forces from so many coalition partners allowed U.S. combat forces to focus their generally superior capabilities in more contested sections of the country. This division of labor served American ends while still ensuring that our partners performed vital work that fully justified their commitment to Iraq's security. These combined operations also strengthened the ties between countries and improved the quality of interoperability between U.S. and coalition troops. Allied support played an important role in stabilizing the situation in Iraq.

This short study also underscores the significant challenges that U.S. Army planners faced in IRAQI FREEDOM in integrating a host of different military partners into U.S. operational plans. Similar issues of working together in a complex military environment will doubtless reoccur in future operations, but the benefits of assembling such coalitions will almost certainly outweigh the problems. The United States cannot fight alone in the current operational environment, and improving the quantity and quality of our interaction with our international partners should continue to be a high priority. I commend this monograph to today's Army to read, gain insight into such combined operations, and reflect on how much support our allies can provide in future military endeavors.

Washington, D.C.   RICHARD W. STEWART
30 September 2011   Chief Historian

Allied Participation in Operation
IRAQI FREEDOM

ALLIED PARTICIPATION IN OPERATION IRAQI FREEDOM

OVERVIEW

From the start of Operation IRAQI FREEDOM in March 2003 until mid-2009, ground troops from thirty-seven countries deployed alongside U.S. forces, with some twenty other countries providing indirect support. Yet U.S. media and public opinion often regarded this "coalition of the willing" with either skepticism or disdain, believing that the United States was engaged in a lonely and, by implication, hopeless cause. On the contrary, many countries provided significant military support to the U.S. Army, performing vital missions to assist combat, intelligence, reconstruction, and support operations. The participation of these coalition partners proved critical to the success of the overall mission.

This monograph examines the achievements and contributions of the thirty-seven allied nations that supplied troops to the U.S.-led coalition in Iraq between 2003 and 2009. The terms *allies*, *coalition forces*, *coalition partners*, and *multinational forces* are used interchangeably to refer to non-U.S. forces that deployed to Iraq under U.S. theater or regional command. These do not include forces deployed to Iraq under the aegis of the United Nations (UN) or the North Atlantic Treaty Organization (NATO), whose troops were not officially part of the U.S. command-and-control structure.

Although the official number is unknown, about sixty nations provided support, both direct and indirect, to the coalition effort in Iraq. Examples of indirect assistance include basing rights, commercial shipping, overflights, and humanitarian aid, among many others. The White House, Department of State, and U.S. Central Command (CENTCOM) all maintained slightly different lists of supporting nations. This monograph, however, focuses on only the thirty-seven nations that furnished direct support in the operation and presents a framework for the more detailed histories to follow.

Current joint-defense doctrine defines multinational operations as those "conducted by forces of two or more nations . . . usually

1

undertaken with[in] the structure of an alliance or coalition." A coalition is defined as "an ad hoc arrangement between two or more nations for a common action," but the doctrine paradoxically adds that "coalitions are formed by different nations with different objectives, usually for a single occasion or for longer cooperation in a narrow sector of interest." No mention is made of multilateral treaty organizations or of unilateral agreements to support allied nations. Nor is any mention made of "host nation support" or formal basing agreements. Thus, although these doctrinal definitions are used throughout this monograph, they do not always provide an adequate basis for understanding the command arrangements adopted by coalition partners in Operation IRAQI FREEDOM.[1]

This monograph also places allied military participation during the operation in context. The IRAQI FREEDOM experience reconfirms the necessity of coalition building in modern warfare, even when U.S. Army and Marine Corps ground forces shoulder the largest burden of the war. Such alliances proved integral for several reasons. First, coalition forces performed vital missions at numerous locations in Iraq, freeing American forces to employ their generally greater combat power in more contested areas of the country. Second, coalition forces often brought unique capabilities to specific operations. For example, Nicaraguan explosive ordnance disposal (EOD) teams demonstrated expertise in dealing with Soviet-era munitions. As a result, they not only destroyed undetonated Soviet explosives expeditiously, but also trained other allied and indigenous Iraqi teams in dealing with such munitions. Moreover, allied presences in Iraq provided diverse approaches to the battlefield, based on each nation's unique experiences and skills. In the realm of civil-military operations, for example, the United Kingdom (U.K.) brought techniques refined in Northern Ireland and in former British colonies; the Dutch used different civil-military procedures based on their own colonial experiences. In many cases allied units tested these techniques and refined or modified them accordingly while making the techniques available to coalition partners, including their American allies. Overall, the great diversity of military participants essentially brought benefits

---

[1] Joint Publication (JP) 3–16, Multinational Operations, 7 March 2007, pp. ix, I-1, Glossary-6. Available online at http://www.fas.org/irp/doddir/dod/jp3-16. pdf.

as well as the expected command-and-control challenges to Operation IRAQI FREEDOM.

In addition to the contributions coalition forces made to the operational and tactical environment in Iraq, their presence had obvious strategic implications that reached far beyond the theater of operations. Multinational operations strengthened and maintained long-standing alliances and reflected a general aura of international cooperation among U.S. allies. They also warded off charges of neocolonialism against the enterprise and imparted a genuine feeling of legitimacy to its aims. Japan's participation in Iraq, its first military deployment since World War II, significantly reinforced its long-existing strategic ties with the United States. A similar bond resulted from El Salvador's comparatively large and lengthy involvement in the war. In addition, partnerships forged in coalition operations helped foster new alliances and friendships. Georgia's decision to send troops to Iraq, for example, created strong ties with the United States and other Western allies, as did the significant contributions of other eastern European and former Communist Bloc nations. Finally, multinational operations helped improve interoperability between U.S. and coalition forces, which will be essential in future combined operations.

The U.S. Central Command and U.S. Army planners faced various problems assembling these coalition forces and establishing a workable command-and-control system. One of the inherent difficulties of military coalition building is accommodating the diverse rules of engagement and national caveats that each allied partner brings to the battlefield. In Operation IRAQI FREEDOM, American doctrine officially recognized that such coalition partners "pick and choose if, when and where they will expend their national blood and treasure."[2] For each nation, the extent of its participation had deep domestic political repercussions, and the identification, appreciation, and accommodation of these constraints was critical for overall success. Thus, in Iraq, some national governments allowed their forces to conduct full-spectrum combat operations, while others significantly restricted the use of their troops. For example, special caveats prohibited some national forces from handling or interrogating detainees, while others restricted the use of

---

[2] JP 3–16, p. xiii.

3

force to self-defense, or specified that troops could operate only with certain coalition partners. Still other coalition governments allowed their forces to operate only on forward operating bases (FOBs) or directed that their forces be used only for specific missions, such as ordnance disposal.

In order to deal with these variations, substantial planning was necessary to match the capabilities of each allied partner with another partner that could complement its capabilities and limitations. In 2004, for example, Japanese engineers, operating under significant limitations that prevented them from conducting self-defense, were partnered with Dutch forces authorized to take part in full-spectrum operations. By 2007, only the United Kingdom, Australia, Estonia, Georgia, Poland, Latvia, Lithuania, and Macedonia authorized their forces to conduct full-spectrum operations. Romania and the Ukraine allowed limited off-base operations, while Albania, Azerbaijan, Bulgaria, the Czech Republic, Mongolia, and Tonga conducted only on-base security and patrols. Matching constraints with needed capabilities in specific geographical areas thus became more of an art than a science, one that had to be practiced regularly as the ground situation changed.

Coalition partners' military capabilities varied considerably and impacted mission availability. Romania, for example, had only military intelligence experts on the ground, while Armenia, Bosnia-Herzegovina, Kazakhstan, and Moldova focused on explosive ordnance disposal. Armenia also provided some transportation and medical support, and Kazakhstan offered water purification services as late as 2007. The Republic of Korea and El Salvador limited themselves to civil-military cooperation, and Japan provided only air support during 2007. Meanwhile, U.S. planners had to factor in the slowly growing capabilities of the Iraqi military and paramilitary forces.

While tailoring operations to take into account these differing capabilities and rules proved difficult, the results offered considerable advantages on the ground. Each coalition member provided additional combat power—*force enablers* in the jargon of the time—that allowed American soldiers to deploy in a more focused manner.

In summary, the coalition forces brought with them additional resources and new capabilities and perspectives that were critical to the evolving war. As the partners became used to working together in a large international force and as they rotated participating forces in and out of the operation, alliances and individual friendships were cemented. The experience of operating in an international environment, which most participants were previously unaccustomed to, became a more normal and accepted part of their expanded professional horizons.

FORMATION OF THE COALITION OF THE WILLING
NOVEMBER 2002–MARCH 2003

In November 2002, President George W. Bush took part in a North Atlantic Treaty Organization summit and announced that Iraqi President Saddam Hussein must disarm or face the consequences at the hands of a United States–led "coalition of the willing." President Bush based his announcement on the provisions of United Nations Security Council Resolution 1441 of 8 November 2002. While the resolution stopped short of authorizing the use of military force against Iraq, it did affirm that Iraq must disarm all of its weapons of mass destruction and long-range missiles. Between Bush's announcement in November 2002 and the commencement of major combat operations in March 2003, the United States built a coalition of allied partners willing to take part, in one form or another, in military operations against Iraq.

By March 2003, the term *Coalition of the Willing* had entered common usage to refer to those countries that supported Operation IRAQI FREEDOM and the U.S. presence in post-invasion Iraq. The initial publicly released list detailing coalition participants in March 2003 contained forty-nine nations. In late 2003, the White House revised this original list, decreasing the number to forty-eight at the request of Costa Rica. Costa Rica's request reflected significant international opposition to military action against Iraq and foreshadowed how difficult it would be to keep such a large and diffuse group together.

In addition to the White House list, the U.S. Department of State and U.S. Senate also published separate sets of coalition members. On 18 March 2003, just prior to the start of Operation IRAQI FREEDOM, the State Department set the coalition number at thirty, while a week into the drive on Baghdad, the Senate put it at fifty. According to the State Department, its number included nations "allowing access, over-flight, or other participation in that way, or they may just have decided they want to be publicly associated with the effort to disarm Iraq." The Senate list included members who had actually deployed military forces, issued a declaration stating the danger posed by Iraq, or had provided diplomatic and strategic support.

In actuality, only four nations sent troops for the initial invasion: the United States, the United Kingdom, Australia, and Poland.[3] The lack of a UN resolution authorizing the use of force deterred others. The United States, Spain, and the United Kingdom introduced such a resolution on 7 March 2003, but the UN Security Council failed to pass it. As a result, some nations, such as Italy and Spain, deployed forces to the theater of operations without taking part in the initial attack.

MAJOR COMBAT OPERATIONS: COALITION FORCES LAND COMPONENT COMMAND, MARCH–MAY 2003

All multinational armed forces that participated in the invasion of Iraq were under the command and control of the commander of CENTCOM, General Tommy R. Franks, although national chains of command remained strong. CENTCOM, a U.S. regional combatant command established in 1983, had planning and operational control of U.S. operations in the Persian Gulf from Egypt to Pakistan. It was a U.S. joint command consisting of land, sea, air, and special operations forces. Each of these components was under the combined-forces command led by General Franks. American and other allied nations' ground forces in CENTCOM served as part of the Coalition Forces Land Component Command (CFLCC), headed by Lt. Gen. David D. McKiernan. CFLCC consisted of two U.S. corps-level headquarters: the U.S. Army V Corps and the I Marine Expeditionary Force. Allied forces, specifically the British 1st Armoured Division, served directly under the I Marine Expeditionary Force during the ensuing operations. Meanwhile, special operations forces of the United Kingdom, Australia, and Poland (and Denmark, according to most reports) became part of the U.S. Combined Forces Special Operations Component

---

[3] Denmark is widely thought to also have deployed special operations forces to assist in the initial attack but has never publicly acknowledged that fact.

6

General Franks *(left)* visits with 101st Airborne Division troops, 7 April 2003.

Chart 1—Coalition Forces Land Component Command and U.S. Combined Forces Special Operations Component Command as of May 2003

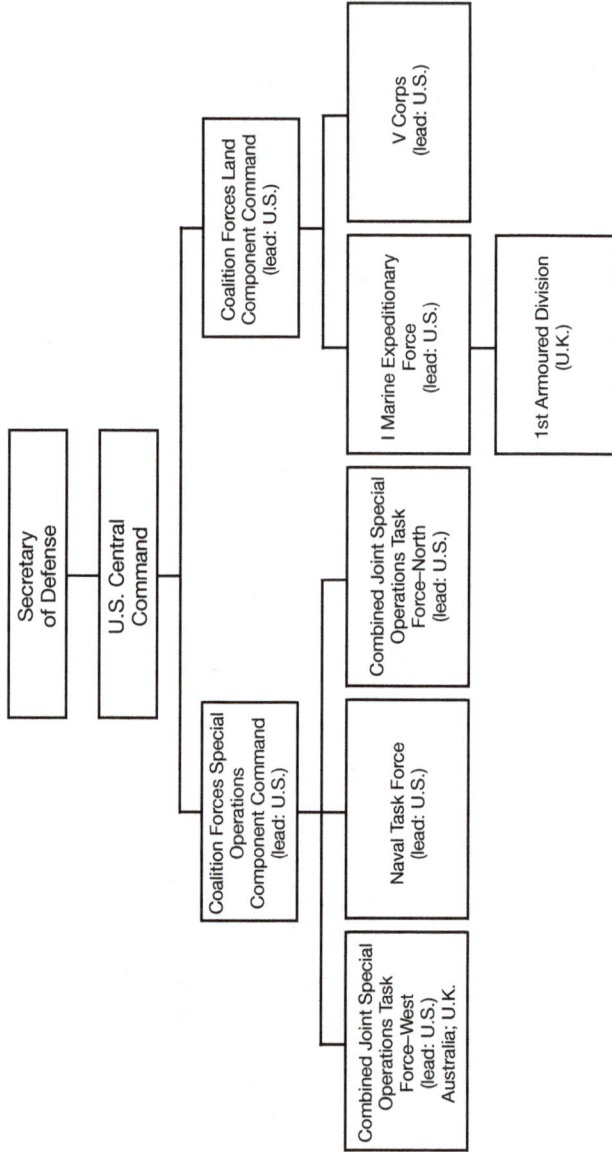

```
                          Secretary
                          of Defense

                          U.S. Central
                           Command

        Coalition Forces Special                Coalition Forces Land
          Operations                            Component Command
       Component Command                          (lead: U.S.)
          (lead: U.S.)

Combined Joint      Naval Task        Combined Joint Special     I Marine              V Corps
Special             Force             Operations Task            Expeditionary         (lead: U.S.)
Operations Task     (lead: U.S.)      Force–North                Force
Force–West                            (lead: U.S.)               (lead: U.S.)
(lead: U.S.)
Australia; U.K.
                                                                 1st Armoured Division
                                                                      (U.K.)
```

Command (CFSOCC), another subordinate headquarters of CENTCOM (*Chart 1*).

Operation IRAQI FREEDOM began on 20 March 2003 at 0534 local time with air strikes followed by a ground offensive into Iraq via Kuwait. While portions of the U.S. V Corps and I Marine Expeditionary Force drove north toward Baghdad, allied forces undertook supporting missions involving both conventional and special operations. On the conventional side, U.K. forces had the task of securing the southeastern city of Al Basrah and the surrounding area. At their first objective, Umm Qasr, a port city near the Iraq-Kuwait border, U.S. marines, along with British forces and Polish commandos, encountered unexpectedly heavy resistance. They gained control of the city after several days of fighting. Immediately, British engineers organized and repaired dock facilities so that supplies could flow into the region directly from ship rather than by road from Kuwait.

British forces had significantly more difficulty capturing Iraq's second-largest city, Al Basrah (*Map 1*). It was not until 6 April, after nearly two weeks of heavy fighting against both regular Iraqi forces and partisans, that they finally secured the urban area. Portions of the U.K. 1st Armoured Division then began advancing north, reaching the town of Al Amarah, some 150 kilometers from Al Basrah, on 9 April to secure a border area that had been hotly contested by Iraq and Iran in the past.

For special operations missions, CFSOCC created three Combined Joint Special Operations Task Forces (CJSOTFs): one to operate in northern Iraq (CJSOTF-N), another for western Iraq (CJSOTF-W), and a third, Naval Task Force, to work the southern coast. Australian and British commandos, operating in CJSOTF-W with U.S. units, secured Scud missile launch sites and blocked potential escape routes into Syria while preventing hostile foreign fighters from entering Iraq in that area. Meanwhile, Polish and U.K. special units in the Naval Task Force conducted operations around Umm Qasr and Al Basrah.

According to intelligence reports prior to the invasion, Iraqi forces had placed explosives on hundreds of oil wells located around Al Basrah and on the Al Faw peninsula. CENTCOM wanted the oil fields seized as rapidly as possible and any planned demolition prevented. Thus, at the start of Operation IRAQI FREEDOM, U.S. marines, joined by British and Polish forces, and supported by Royal Navy, Polish Navy, and Royal Australian Navy warships, made an amphibious assault on the

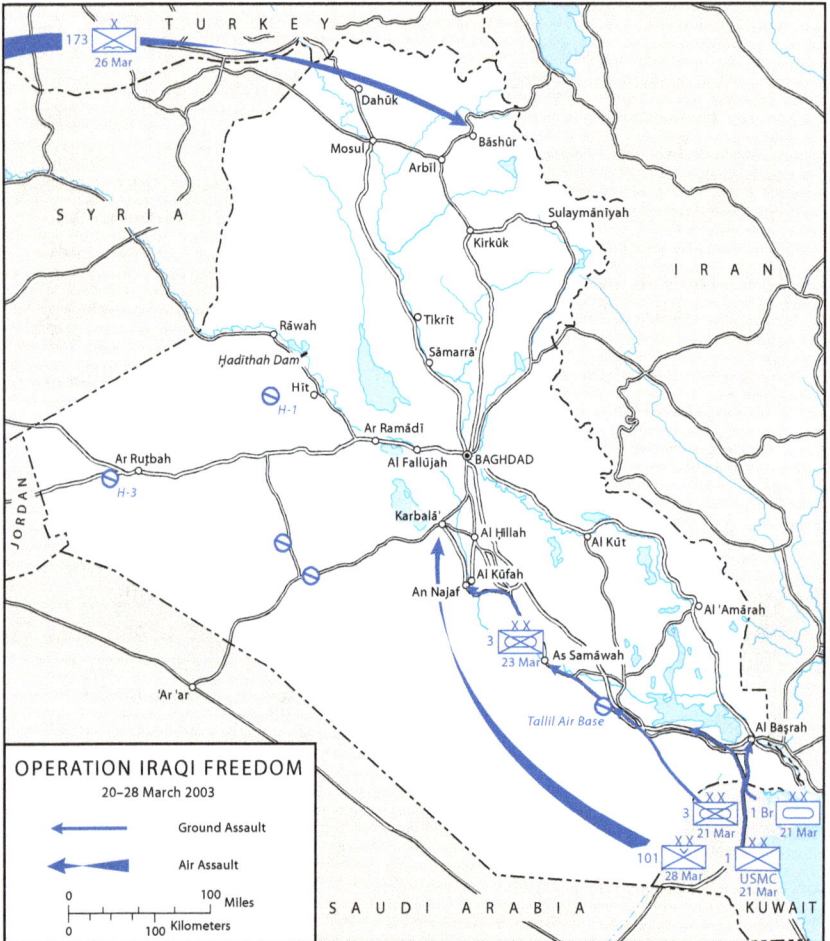

OPERATION IRAQI FREEDOM
20–28 March 2003

Ground Assault

Air Assault

0 — 100 Miles

0 — 100 Kilometers

Map 1

Al Faw peninsula. Another British force, the 16th Air Assault Brigade, secured the oil fields in southern Iraq around Rumaylah, while Polish commandos captured offshore oil platforms near Umm Qasr. These forces completed all tasks successfully.

Shortly after the initial invasion, Spanish soldiers moved into southern Iraq from positions in Kuwait. Although they later served in a combat role alongside British and Australian forces, their first task was primarily humanitarian assistance behind the allied lines

10

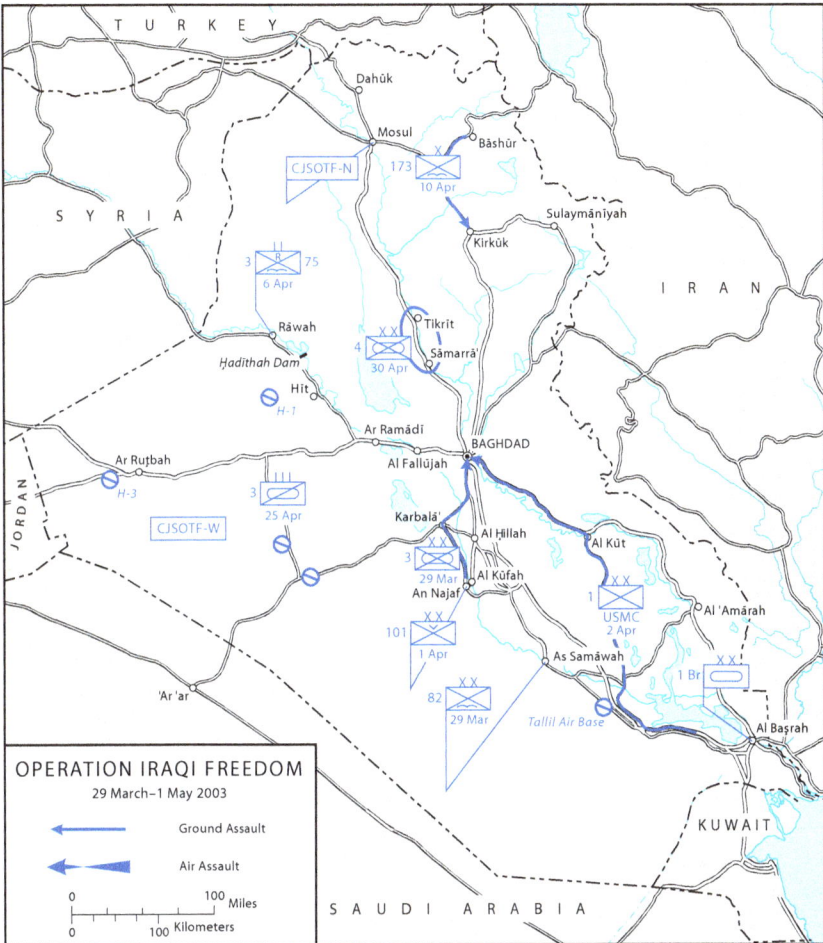

OPERATION IRAQI FREEDOM
29 March–1 May 2003

Ground Assault

Air Assault

0 — 100 Miles
0 — 100 Kilometers

Map labels: TURKEY, Dahūk, Mosul, Bāshūr, CJSOTF-N, 173, 10 Apr, Sulaymānīyah, Kirkūk, SYRIA, 3 R 75, 6 Apr, Rāwah, Tikrīt, 4, Sāmarrā', 30 Apr, Ḥadīthah Dam, Hit, H-1, Ar Ramādī, BAGHDAD, Ar Ruṭbah, Al Fallūjah, IRAN, H-3, 3, 25 Apr, CJSOTF-W, Karbalā', Al Ḥillah, Al Kūt, 3, 29 Mar, Al Kūfah, An Najaf, 1 USMC, 2 Apr, Al 'Amārah, 101, 1 Apr, As Samāwah, 'Ar 'ar, 82, 29 Mar, Tallil Air Base, 1 Br, Al Baṣrah, KUWAIT, SAUDI ARABIA, JORDAN

Map 2

of advance. A number of other nations sent troops to Iraq during this phase; Albania, Czech Republic, Dominican Republic, El Salvador, Republic of Korea, Ukraine, and Mongolia all deployed soldiers to Iraq from late March through April 2003 (*Map 2*). By 31 April 2003, a total of eleven allied partners had committed nearly fifty thousand troops to the coalition effort.

Following coalition successes in southern, western, and northern Iraq, and the capture of Baghdad on 9 April 2003,

General Sanchez (*right*) walks with Secretary of Defense Donald H. Rumsfeld, who arrived at Baghdad International Airport, 6 December 2003.

President Bush declared an end to major combat operations on 1 May 2003. As a result, the number of allied troops in Iraq soon fell to twenty-one thousand; however, the number of allied nations participating in Iraq grew as the U.S.-led coalition transitioned to Phase IV occupation through the remainder of the year.[4]

Combined Joint Task Force–7, June 2003–May 2004

On 14 June 2003, after the declared end of major combat operations, Combined Joint Task Force–7 (CJTF-7) replaced CFLCC as the strategic, operational, and tactical headquarters for all ground forces in theater (Iraq and Kuwait). CFLCC,

---

[4] Using the standard planning practice, the CENTCOM staff issued planning guidance specifying four phases of military operations for Operation Iraqi Freedom: Phase I, to deter and engage the enemy; Phase II, to seize the initiative; Phase III, to conduct decisive operations; and Phase IV, to transition to peace.

however, tentatively remained the primary logisticical hub for the theater of operations in Kuwait. The primary element of CJTF-7, the U.S. Army's V Corps, was commanded first by Lt. Gen. William S. Wallace and then, beginning in July, by Lt. Gen. Ricardo Sanchez. Generals Wallace and Sanchez commanded both CJTF-7 and V Corps.

During the beginning of Phase IV operations, CJTF-7's main task was to secure the establishment of an interim Iraqi government, the Iraqi Governing Council (IGC), and the Coalition Provisional Authority (CPA), the latter of which exercised executive, legislative, and judicial authority over the IGC from its establishment on 21 April 2003 until its dissolution on 28 June 2004. The CPA replaced the short-lived Office of Reconstruction and Humanitarian Assistance, headed by Lt. Gen. Jay Garner (USA, Ret.), when former Ambassador L. Paul Bremer took control of the CPA on 11 May 2003. Soon after, CJTF-7 assumed responsibility for organizing, training, and certifying a newly created indigenous security force when Ambassador Bremer mandated the disbandment of all Iraqi armed forces on 23 May.

To perform its diverse missions, CJTF-7 also had at least limited operational control of all forces within Iraq, including all multinational forces. After 1 May 2003, twenty countries deployed contingents to Iraq: Azerbaijan, Bulgaria, Estonia, Georgia, Honduras, Hungary, Italy, Kazakhstan, Latvia, Lithuania, Macedonia, Moldova, the Netherlands, New Zealand, Nicaragua, Norway, Philippines, Romania, Slovakia, and Thailand. The influx brought the total number of allied partners in Iraq to thirty-three and increased the number of non-U.S. troops in Iraq to twenty-five thousand. Key to this second expansion was UN Security Council Resolution 1511, passed on 16 October, which officially authorized the creation of a multinational security force in Iraq and urged UN member states to contribute to the force and the reconstruction of Iraq. More nations sent troops after the UN approved the resolution because it provided international legitimacy to the effort.

For command-and-control purposes, CJTF-7 divided Iraq into six divisional areas of responsibility: Multi-National Division–North (MND-N), Multi-National Division–North Central (MND-NC), Multi-National Division–Baghdad (MND-B), Multi-National Division–West (MND-W), Multi-National Division–Center-South

Retired General Garner *(right)* joins Secretary of Defense Rumsfeld for a briefing at the Pentagon, 18 June 2003.

(MND-CS), and Multi-National Division–Southeast (MND-SE) (*Chart 2*). Later in 2004, MND-N split into Multi-National Force–Northwest (MNF-NW) and Multi-National Division–Northeast (MND-NE).[5]

Allied partners played the major role in several of these multinational divisions. (*See Map 3.*) The British, who provided the

---

[5] Future changes to the Multi-National Divisions included the following: In 2006 MND-NC merged with MND-N. In 2007, with the surge of forces, MND-C (Center) formed. In 2008 MND-NE merged with MND-N, and MND-CS merged with MND-C. In 2009 MND-C merged with MND-SE, with the latter becoming MND-S later that year. In January 2010, Multi-National Divisions North, Baghdad, and South transitioned to U.S. Divisions North, Center, and South.

Chart 2—Combined Joint Task Force–7 as of June 2003

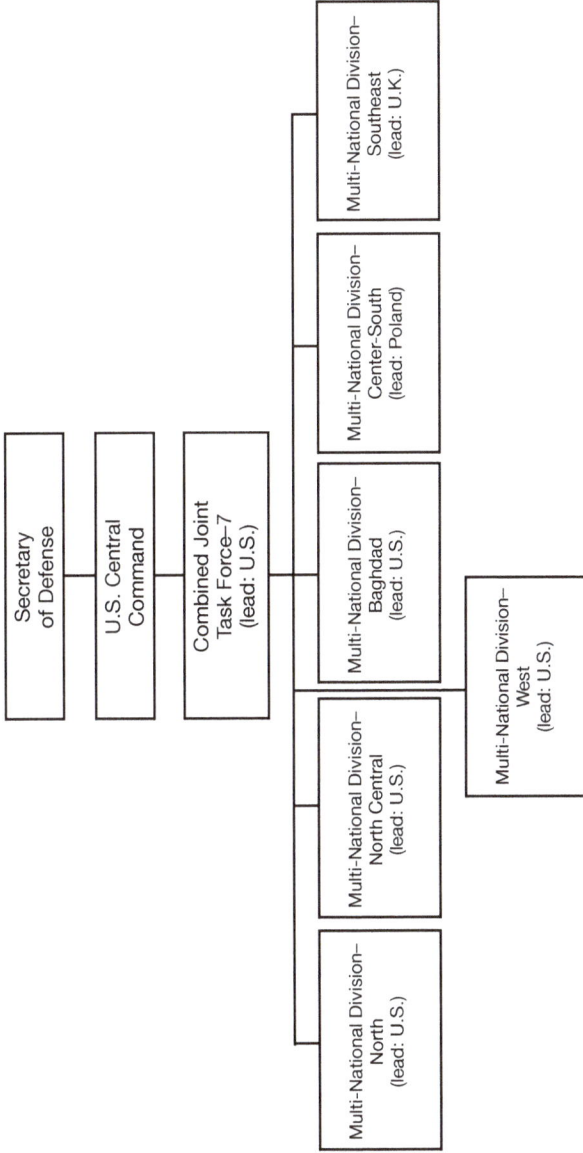

```
                    ┌─────────────────┐
                    │   Secretary     │
                    │  of Defense     │
                    └────────┬────────┘
                             │
                    ┌────────┴────────┐
                    │  U.S. Central   │
                    │    Command      │
                    └────────┬────────┘
                             │
                    ┌────────┴────────┐
                    │ Combined Joint  │
                    │  Task Force–7   │
                    │   (lead: U.S.)  │
                    └────────┬────────┘
                             │
   ┌──────────────┬──────────┼──────────────┬──────────────────┐
   │              │          │              │                  │
┌──┴───────┐ ┌────┴─────┐ ┌──┴───────┐ ┌────┴──────┐ ┌─────────┴────┐
│Multi-    │ │Multi-    │ │Multi-    │ │Multi-     │ │Multi-        │
│National  │ │National  │ │National  │ │National   │ │National      │
│Division– │ │Division– │ │Division– │ │Division-  │ │Division–     │
│North     │ │North     │ │Baghdad   │ │Center-    │ │Southeast     │
│(lead:    │ │Central   │ │(lead:    │ │South      │ │(lead: U.K.)  │
│U.S.)     │ │(lead:    │ │U.S.)     │ │(lead:     │ │              │
│          │ │U.S.)     │ │    │     │ │Poland)    │ │              │
└──────────┘ └──────────┘ └────┼─────┘ └───────────┘ └──────────────┘
                               │
                     ┌─────────┴────────┐
                     │  Multi-National  │
                     │  Division–       │
                     │  West            │
                     │  (lead: U.S.)    │
                     └──────────────────┘
```

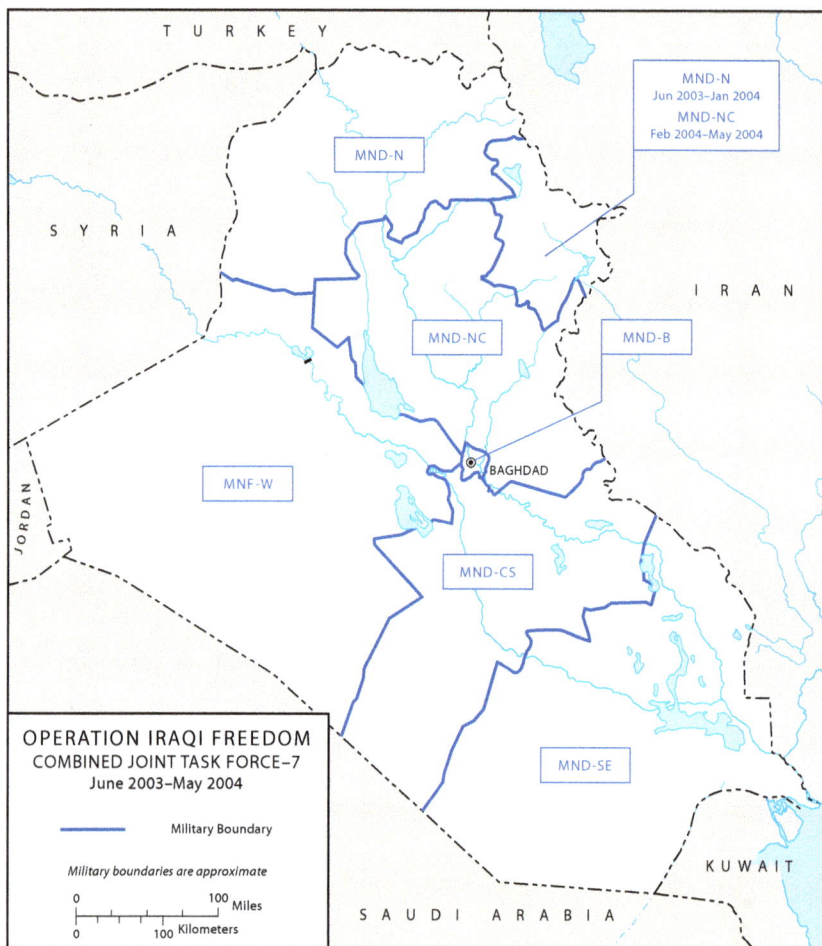

OPERATION IRAQI FREEDOM
COMBINED JOINT TASK FORCE–7
June 2003–May 2004

Military Boundary

Military boundaries are approximate

Map 3

second-largest contingent of forces in Iraq, took the lead in MND-SE and provided command and control for a large number of coalition forces in southeastern Iraq, including those from Italy, Australia, Romania, Denmark, Portugal, Czech Republic, and Lithuania. Poland, the third-largest troop contributor, took the lead in MND-CS, which included troops from the Ukraine; the so-called Plus Ultra Brigade (MNB PU), also known as the Brigada

Hispanoamericana, composed of troops from Spain, Dominican Republic, El Salvador, Honduras, and Nicaragua; as well as forces from Kazakhstan, Latvia, and Mongolia.[6] MND-NE, under the lead of the Republic of Korea, stood up in September 2004 after Seoul agreed to significantly increase its troop strength and undertake reconstruction and humanitarian efforts in northern Iraq.

On the operational front, allied forces experienced both successes and problems during CJTF-7's tenure. During the summer and fall of 2003, coalition forces placed considerable emphasis on tracking and eliminating the remaining leaders of Saddam Hussein's regime. The most notable event in this initial hunt occurred on 22 July when Saddam Hussein's two sons, Uday and Qusay, died in a firefight with U.S. forces. In all, CJTF-7 captured or killed some two hundred leaders of the regime relatively quickly. Coalition forces did not capture the fugitive dictator until December 2003.

During this same period an insurgency against the coalition occupation began to take root. Ambassador Bremer's decision to demobilize and institute a de-Ba'athification process greatly contributed to the rise of the insurgency. Disgruntled Iraqis began attacking coalition forces. Simple ambushes and snipers, mines, and indirect fire evolved into more sophisticated improvised explosive devices (IEDs), suicide bombings, larger and better-led hit-and-run attacks, and general acts of terror. The attackers often targeted Iraqi civilians of different ethnicities, tribal groups, religious sects, as well as coalition forces. As a result, counterinsurgency became a growing part of coalition Phase IV operations in Iraq.

In April 2004, CJTF-7 commanded 162,000 ground troops in Iraq, of which approximately 25,000 were coalition soldiers and a Japanese contingent that deployed to Iraq in February 2004. A number of nations, however, withdrew their forces during this time. Nicaraguan forces left Iraq in February 2004 due to funding constraints. Then, in the wake of the bombing of the Madrid commuter trains in March 2004 by terrorists, Spain withdrew its forces, as did the rest of the Plus Ultra Brigade except El Salvador. These withdrawals resulted in the number of allied nations with troops in Iraq falling to thirty. In part as

---

[6] The term *Plus Ultra*, which is Latin for "further beyond," is the national motto of Spain.

General Casey arrives on Forward Operating Base Dagger in Tikrit, 8 August 2006.

a response to the deteriorating situation, on 15 May, less than eleven months after CJTF-7 stood up, CENTCOM replaced it with two new commands: Multi-National Force–Iraq (MNF-I) and Multi-National Corps–Iraq (MNCI). General Sanchez, the CJTF-7 commander, led MNF-I until 1 July, when General George W. Casey replaced him, while Lt. Gen. Thomas F. Metz took command of MNC-I.

MULTI-NATIONAL FORCE–IRAQ, MAY 2004–JULY 2009

The deployment of allied forces under MNF-I falls into five stages. Each is tied to alterations in MNF-I multinational division boundaries or significant changes in U.S. or coalition deployments. Stage 1 (May 2004–May 2005) encompasses the creation and initial year of the MNF-I headquarters; Stage 2 (June 2005–December 2006) tracks several changes in the multinational division command-and-control structure; Stage 3 (January 2007–July 2008) covers both the U.S. troop surge and the increased numbers of well-trained and well-equipped Iraqi forces; Stage 4 (August–December 2008) reflects continuation of the surge but with most remaining coalition partners withdrawing after the expiration of the UN mandate in Iraq; and, finally, Stage 5 (January–July 2009) refers to the period when the few remaining allied forces permanently left Iraq in accordance with agreements with Baghdad.

Multi-National Force–Iraq: Stage 1, May 2004–May 2005

Coalition leaders created MNF-I to overcome several fundamental problems with the CJTF-7 organization. CJTF-7 had been responsible for all aspects of strategic, operational, and tactical control of all forces in Iraq, a mission too broad for a single organization. In place of CJTF-7, MNF-I provided theater-level strategic and operational planning while leaving three subordinate commands to handle more specific tasks: Multi-National Corps–Iraq planned and conducted day-to-day tactical operations; the Multi-National Security Transition Command–Iraq (MNSTC-I) coordinated coalition efforts to train and equip the new Iraqi Security Forces (ISF); and the U.S. Army Corps of Engineers Gulf Region Division coordinated and supervised American reconstruction efforts in Iraq. This reorganization did not, however, solve the problem of unity of command.

While MNSTC-I technically oversaw all coalition ISF training efforts, additional trainers came from the United Nations Assistance Mission for Iraq (UNAMI), formed by UN Security Council Resolution 1500 on 14 August 2003, and from the NATO Training Mission–Iraq (NTM-I) at the request of the Iraqi Interim Government under the provisions of UN Security Council Resolution 1546. Neither was part of the MNF-I force structure or under MNF-I command and control. In addition, many other trainers provided advice and assistance to the fledgling Iraqi governmental organization, including its police and security forces, while other agencies provided general assistance to the Iraqi economy and social infrastructure. Bringing the entire allied

19

Chart 3—Multi-National Force–Iraq as of June 2004

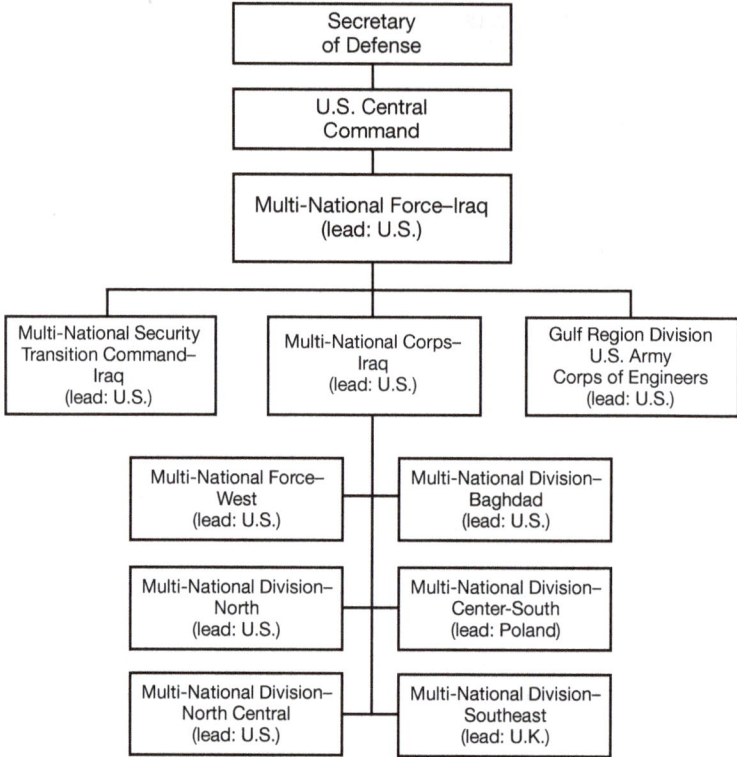

```
                    ┌─────────────────────┐
                    │      Secretary      │
                    │     of Defense      │
                    └─────────────────────┘
                              │
                    ┌─────────────────────┐
                    │    U.S. Central     │
                    │      Command        │
                    └─────────────────────┘
                              │
                    ┌─────────────────────┐
                    │ Multi-National Force–Iraq │
                    │     (lead: U.S.)    │
                    └─────────────────────┘
                              │
     ┌────────────────────────┼────────────────────────┐
┌──────────────────┐  ┌──────────────────┐  ┌──────────────────┐
│ Multi-National   │  │ Multi-National   │  │ Gulf Region      │
│ Security         │  │ Corps–           │  │ Division         │
│ Transition       │  │ Iraq             │  │ U.S. Army        │
│ Command–Iraq     │  │ (lead: U.S.)     │  │ Corps of         │
│ (lead: U.S.)     │  │                  │  │ Engineers        │
└──────────────────┘  └──────────────────┘  │ (lead: U.S.)     │
                                             └──────────────────┘
```

Multi-National Force–West (lead: U.S.)	Multi-National Division–Baghdad (lead: U.S.)
Multi-National Division–North (lead: U.S.)	Multi-National Division–Center-South (lead: Poland)
Multi-National Division–North Central (lead: U.S.)	Multi-National Division–Southeast (lead: U.K.)

assistance mission in Iraq under one roof remained an elusive goal throughout this period.

Shortly after standing up, MNF-I adjusted the boundaries of the multinational division areas of responsibility. The basic multinational division structure, however, remained relatively unchanged with MND-N, MND-NC, MNF-W, MND-B, MND-CS, and MND-SE (*Chart 3*). The only significant alteration was the creation of MND-NE, under the control of the Republic of Korea, from the MND-N area of responsibility. As a result, MNF-I truncated MND-N and renamed it Multi-National Force–Northwest (MNF-NW) in September 2004. The United Kingdom

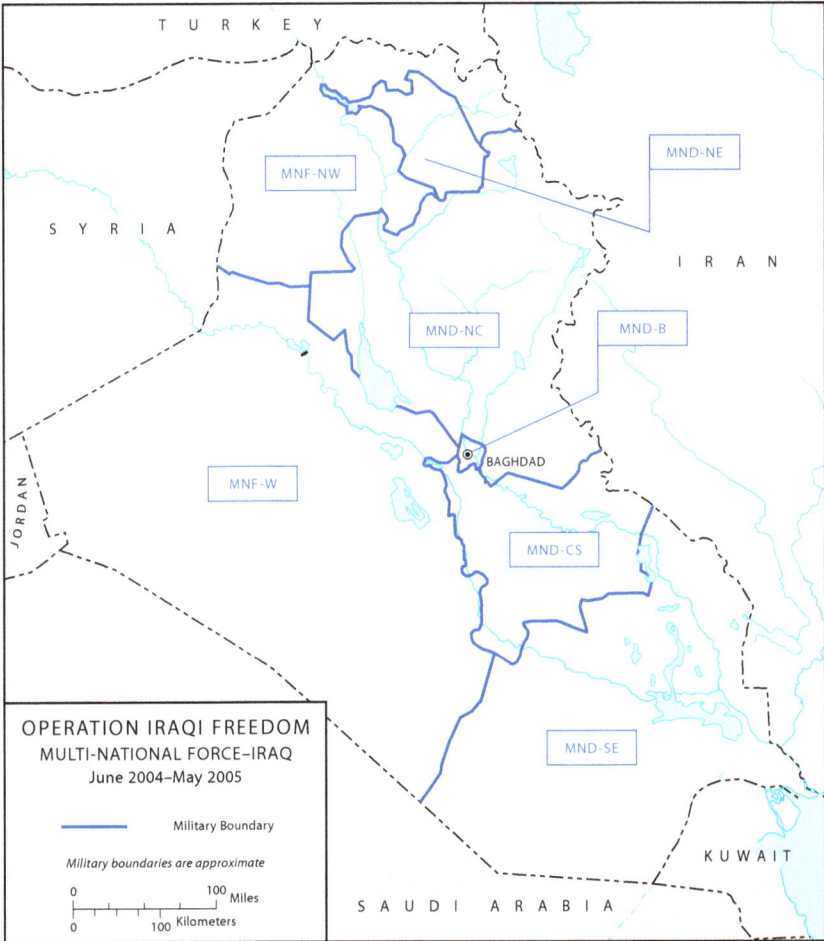

OPERATION IRAQI FREEDOM
MULTI-NATIONAL FORCE–IRAQ
June 2004–May 2005

Military Boundary

Military boundaries are approximate

0   100 Miles
0   100 Kilometers

Map 4

and Poland continued to provide the lead in MND-SE and MND-CS, respectively (*Map 4*).

During its first full year, MNF-I saw alterations in the allied nations serving in Iraq. Some nations withdrew their military forces after their mandates had expired or after they had accomplished the missions that their national governments had set. Norway and the Philippines withdrew their forces in June and July 2004,

21

respectively—the Philippines doing so in direct response to the kidnapping of a Filipino contractor by insurgents. After providing troops for one year, Thailand removed its forces in September 2004 as did New Zealand in October 2004. Other members that pulled their troops out of Iraq included Hungary in December 2004, and Portugal and the Netherlands in February and March 2005, respectively. For the Dutch, their mission ended after the Coalition returned the province of Al Muthanna to Iraqi control and made it responsible for its own security.

Meanwhile, a few smaller nations added their units to the mix. Tonga deployed a small contingent to Iraq in June 2004, and Armenia deployed its first forces, fifty personnel, in February 2005. Other nations added or subtracted from their deployed forces so that by May 2005, MNF-I maintained command and control over twenty-three thousand non-U.S. coalition soldiers. This was only two thousand fewer soldiers than before CJTF-7 stood down the previous year. While the number of allied partners in the Coalition decreased to twenty-five by May 2005, many of the remaining allied contingents actually increased in size.

Operationally, MNF-I directed major offensives against insurgent strongholds in Al Fallujah and Mosul in November 2004. These offensives were conducted largely by U.S. troops, while allied forces focused on counterinsurgency and reconstruction. In conjunction with these military operations, the Iraqi indigenous administrations exercised limited sovereignty by supervising a series of elections throughout the year. One election, on 30 January 2005, selected representatives for a newly formed national assembly, which crafted and ratified a new constitution on 15 October 2005. The assembly also supervised a general election on 15 December 2005 that created a permanent 275-member Iraqi Council of Representatives.

Multi-National Force–Iraq: Stage 2, June 2005–December 2006

As of June 2005, General Casey, the MNF-I commander, had made no additional alterations to the existing multinational division structure (*Chart 4*). Only later in 2005 did he eventually disband MND-NC and MNF-NW to re-create MND-N. There was also little fluctuation in the coalition command system or in the allied contributions to MNF-I during the next eighteen months. The number of participating coalition partners decreased to twenty-two after Bulgaria and Ukraine terminated their missions

Chart 4—Multi-National Force–Iraq as of June 2005

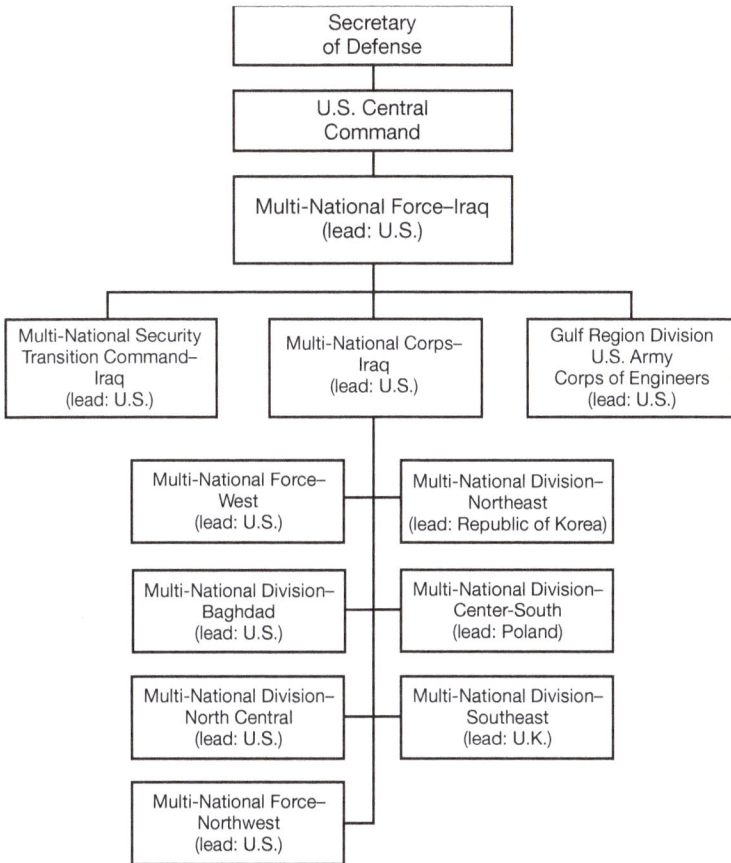

```
                    ┌─────────────────────┐
                    │     Secretary       │
                    │    of Defense       │
                    └─────────────────────┘
                    ┌─────────────────────┐
                    │   U.S. Central      │
                    │    Command          │
                    └─────────────────────┘
                    ┌─────────────────────┐
                    │ Multi-National Force–Iraq │
                    │     (lead: U.S.)    │
                    └─────────────────────┘
```

| Multi-National Security Transition Command–Iraq (lead: U.S.) | Multi-National Corps–Iraq (lead: U.S.) | Gulf Region Division U.S. Army Corps of Engineers (lead: U.S.) |

Multi-National Force–West (lead: U.S.)	Multi-National Division–Northeast (lead: Republic of Korea)
Multi-National Division–Baghdad (lead: U.S.)	Multi-National Division–Center-South (lead: Poland)
Multi-National Division–North Central (lead: U.S.)	Multi-National Division–Southeast (lead: U.K.)
Multi-National Force–Northwest (lead: U.S.)	

in December 2005, and Japan and Italy ended their deployments in July 2006. Still, the actual number of coalition forces remained fairly consistent: between 18,000 and 23,000 soldiers. It was not until December 2006 that the numbers of non-U.S. soldiers decreased to 15,000.

At the strategic and operational level, MNF-I was confronted with an ever-growing number of insurgent incidents in 2005 and 2006. In 2005 alone, 34,131 insurgent attacks took place, up significantly from 26,496 in 2004, reaching a crescendo on 22 February 2006 with the bombing of the al-Askari Mosque in Samarra. The attack on one of the holiest Shi'ite sites caused no injuries but led to a wave of retaliatory violence throughout the region and across Iraq. To many observers, the country seemed poised to erupt in open civil strife based on religious, ethnic, and tribal differences. In response, MNF-I began planning for a significant surge of U.S. and allied forces in 2007.

Multi-National Force–Iraq: Stage 3, January 2007–July 2008 ("The Surge")

After significant study and debate in late 2006, President Bush decided to dramatically increase U.S. troop levels in Iraq, announcing his decision during a televised presidential address on 10 January 2007. The subsequent military "surge" included sending five additional Army brigades, some twenty thousand troops, to Iraq beginning in January 2007. To facilitate the troop surge, MNF-I created a new area of responsibility called Multi-National Division–Center (MND-C), which maintained responsibility from the outskirts of Baghdad to the Kuwait border and relieved MND-B forces of their role outside the city. In the old CJTF-7 and earlier MNF-I organizations, MND-B forces had been responsible not only for Baghdad proper but also the vast stretch of territory to the south; this change in roles enabled MND-B to focus more closely on the vital capital area (*Chart 5*).

In February, MNF-I in conjunction with Iraqi forces launched a major operation to secure Baghdad, Operation Fardh al-Qanoon ("imposing law"), while coalition forces launched additional counterinsurgency operations across Iraq. For example, Operation Phantom Thunder, which began on 16 June 2007, targeted insurgents in Diyala and Al Anbar Provinces. Operations Phantom Strike and Phantom Phoenix targeted other areas aggressively.

During the surge, MNF-I counterinsurgency strategy changed significantly. New tactics were adopted to reduce civilian casualties, and commanders were encouraged to develop closer relations with local leaders and citizens. Such efforts required not only a more precise employment of combat power, but also more positive interaction between the local populace and small U.S. and allied units.

While the surge and the new tactics did not appear to decrease violence at first, the situation began to turn by September 2007. More security forces, better and closer relations with the people,

24

Chart 5—Multi-National Force–Iraq as of June 2007

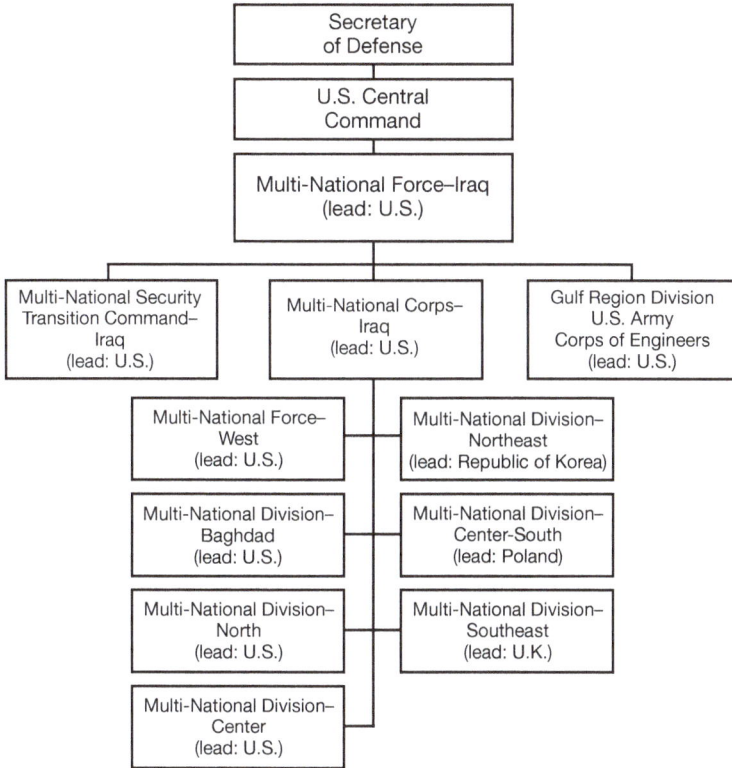

```
                    ┌─────────────────────┐
                    │     Secretary       │
                    │    of Defense       │
                    └─────────────────────┘
                    ┌─────────────────────┐
                    │    U.S. Central     │
                    │     Command         │
                    └─────────────────────┘
                    ┌─────────────────────┐
                    │ Multi-National Force–Iraq │
                    │     (lead: U.S.)    │
                    └─────────────────────┘
```

| Multi-National Security Transition Command–Iraq (lead: U.S.) | Multi-National Corps–Iraq (lead: U.S.) | Gulf Region Division U.S. Army Corps of Engineers (lead: U.S.) |

Multi-National Force–West (lead: U.S.)	Multi-National Division–Northeast (lead: Republic of Korea)
Multi-National Division–Baghdad (lead: U.S.)	Multi-National Division–Center-South (lead: Poland)
Multi-National Division–North (lead: U.S.)	Multi-National Division–Southeast (lead: U.K.)
Multi-National Division–Center (lead: U.S.)	

a growing number of Sunnis taking up arms against al-Qaeda, and more aggressive operations against insurgent strongholds slowly began to reduce at least the outward manifestations of the insurgency. These trends continued into 2008. In addition, Iraqi Security Forces began assuming larger roles in combined allied-Iraqi operations. As the number of trained, well-equipped, and experienced ISF personnel increased, the number of required coalition forces in any one area proportionally decreased.

During this period only two nations ended their deployments: Slovakia in February 2007 and Denmark in August 2007. Tonga redeployed forces to Iraq in 2007. Through much of that year, the number of allied soldiers in Iraq held at about 12,000. When U.S. forces began to draw down following the surge in early 2008,

General David H. Petraeus *(second from left)* arrives at Camp Gannon, 6 September 2008. General Petraeus was the Multi-National Force–Iraq commander in February 2007–September 2008.

coalition forces also decreased, and by June 2008 only about 9,700 allied troops remained in Iraq.

MULTI-NATIONAL FORCE–IRAQ: STAGE 4, AUGUST–DECEMBER 2008

The expiration of UN Security Council Resolution 1790 posed a significant obstacle to further coalition participation in Operation IRAQI FREEDOM. Originally passed on 18 December 2007, the resolution extended the mandate for multinational forces to remain in Iraq until 31 December 2008. In fact, every year since 2004, the UN Security Council renewed the mandate for multinational forces at the request of the Iraqi government. In 2008, however, Iraqi officials signaled their desire to end the mandate. As a result, the resolution expired and a majority of the allied forces in Iraq withdrew by the end of that year, including a number of nations that had promised to remain in Iraq as long as U.S. forces stayed in country. National contingents from Albania, Armenia, Azerbaijan, Bosnia-Herzegovina, Bulgaria, the Czech Republic, Kazakhstan, Latvia, Lithuania, Macedonia, Moldova, Mongolia, Poland, Republic of Korea, and Tonga all left the country. Georgia also unexpectedly withdrew its forces in August 2008 because of its short but intense conflict with Russia (*Map 5*).

In a final agreement approved by the Iraqi parliament on 4 December 2008, the United States and Iraq entered into a new status-of-forces agreement. The "Agreement between the United States of America and the Republic of Iraq on the Withdrawal of United States Forces from Iraq and the Organization of Their Activities during Their Temporary Presence in Iraq" allowed U.S. forces to remain in Iraq past the expiration of the UN mandate, with caveats. It required U.S. combat forces to relocate away from Iraqi urban areas by 30 June 2009 and the complete withdrawal of U.S. forces by 31 December 2011. The agreement also significantly limited the conduct of U.S. military operations in Iraq.

On 18 December, the Iraqi government signed bilateral agreements with other allied countries regarding the presence of their combat troops past the 31 December 2008 expiration of the UN mandate. Subsequently it made such agreements with Australia, El Salvador, Estonia, Romania, and the United Kingdom, allowing their forces to remain into 2009. By the end of 2008, the number of allied nations with forces in Iraq had thus decreased from twenty-one to five. In terms of troop numbers, allied personnel fell from 9,734 in July 2008 to about 5,000 by 31 December 2008. The withdrawal also required significant realignment of the MND structure. MNF-I dissolved MND-NE in late 2008 and MND-CS in early 2009, followed by the dissolution of MND-SE after the British withdrew from their area of responsibility in May 2009 (although the last British troops did not depart until 28 July). This left only MND-N, MND-C, MND-S, and MNF-W.

Multi-National Force–Iraq: Stage 5, January–July 2009

In the first seven months of 2009, all the remaining allied nations withdrew their forces from Iraq in accordance with their

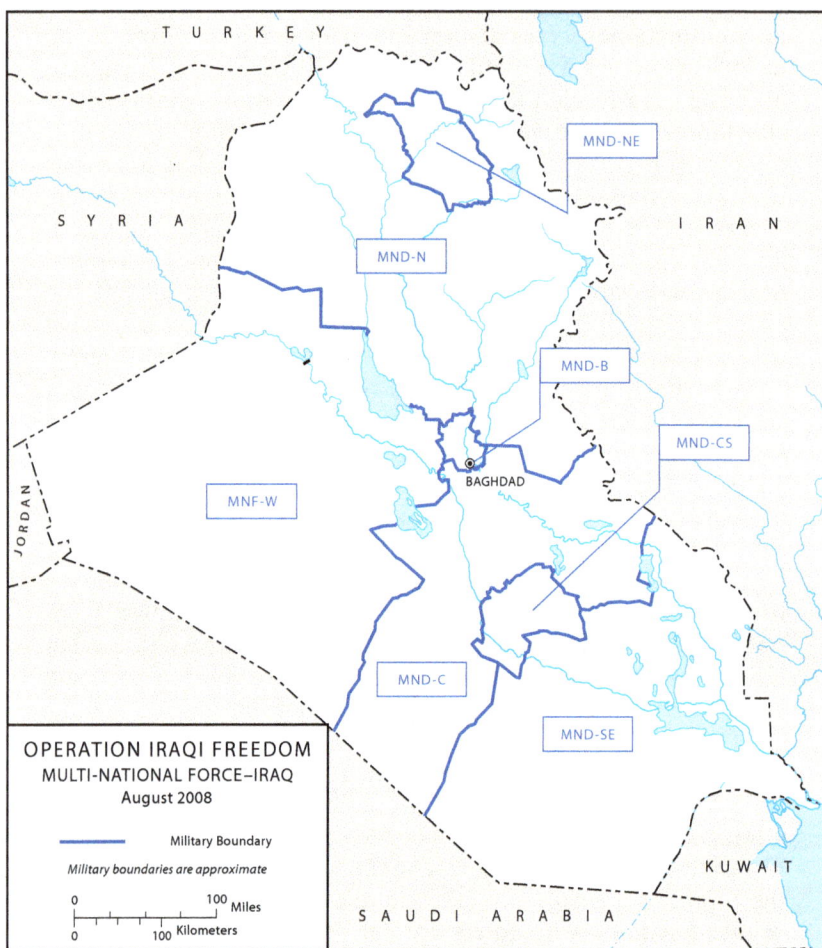

OPERATION IRAQI FREEDOM
MULTI-NATIONAL FORCE–IRAQ
August 2008

——— Military Boundary

Military boundaries are approximate

0   100 Miles
0   100 Kilometers

Map 5

bilateral agreements with the Iraqi government. (*See Chart 6.*) The last Salvadoran troops departed Iraq on 22 January 2009; the forces had been deployed in Iraq since August 2003 and had remained even after the Spanish-led Plus Ultra Brigade left in early 2004. Estonia, which had personnel deployed to Iraq since June 2003, terminated its mission on 7 February 2009.

Romania, which originally had 730 soldiers deployed to Iraq, had pledged to stay with U.S. forces through 2011. When it became

28

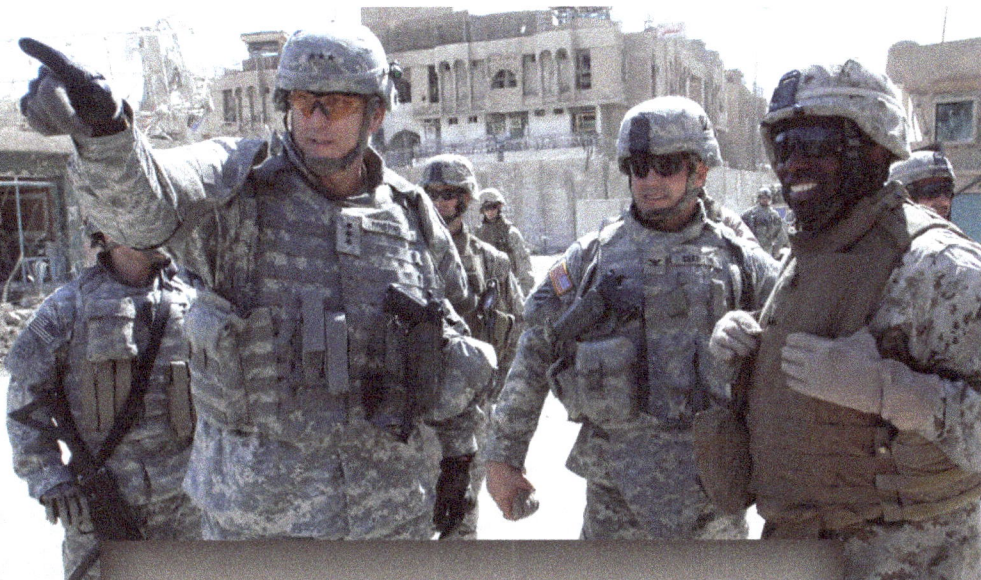

Lt. Gen. Raymond T. Odierno *(second from left)* points out a location in downtown Ar Ramadi, 25 June 2007. General Odierno was the Multi-National Force–Iraq commander in September 2008–December 2009.

clear, however, that the Iraqi government would not approve coalition missions beyond 31 July 2009, Bucharest officially terminated its mission on 4 June 2009, and the last Romanian troops left Iraq on 23 July. (*See Map 6.*)

An agreement between Canberra and Baghdad led to Australian forces withdrawing from Iraq on 28 July 2009. They previously served in the Al Muthanna Task Group, which replaced outgoing Dutch forces in Al Muthanna Province, in Overwatch Battle Group–West, and in specialist missions. The only remaining members of its security detachment were responsible for protecting the Australian Embassy in Baghdad.

Officially, British combat operations in Iraq ended on 30 April 2009. The bulk of the remaining U.K. contingent, still some 3,300 strong, withdrew by the end of May 2009, with the last British forces leaving the country on 28 July. A number of British military personnel, however, relocated to Kuwait and

Chart 6—Multi-National Force–Iraq as of June 2009

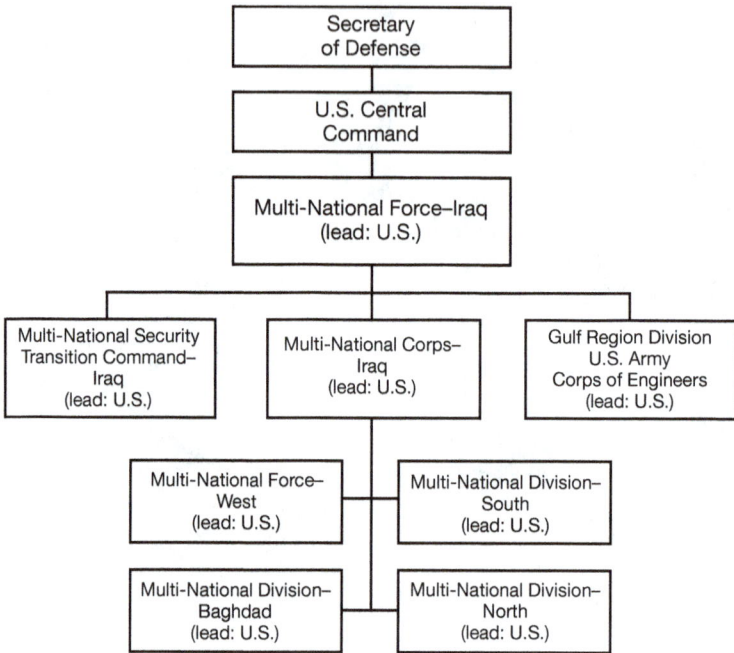

```
                        ┌─────────────────────┐
                        │     Secretary       │
                        │    of Defense       │
                        └─────────────────────┘
                                  │
                        ┌─────────────────────┐
                        │   U.S. Central      │
                        │    Command          │
                        └─────────────────────┘
                                  │
                        ┌─────────────────────┐
                        │ Multi-National Force–Iraq │
                        │     (lead: U.S.)    │
                        └─────────────────────┘
```

| Multi-National Security Transition Command–Iraq (lead: U.S.) | Multi-National Corps–Iraq (lead: U.S.) | Gulf Region Division U.S. Army Corps of Engineers (lead: U.S.) |

| Multi-National Force–West (lead: U.S.) | Multi-National Division–South (lead: U.S.) |

| Multi-National Division–Baghdad (lead: U.S.) | Multi-National Division–North (lead: U.S.) |

remained in theater at the request of the Kuwaiti government. As a result of the British withdrawal, MND-SE was dissolved and replaced with a new Multi-National Division–South (MND-S), which was created by merging MND-CS and MND-SE. The withdrawals left only MND-B, MND-N, MND-S, and MNF-W intact, all manned by U.S. forces.

ANALYSIS

For over five critical years, coalition military forces held down the southern region of Iraq, helped minimize the spread of the insurgency to that key area, and thereby protected the logistical base of the central, western, and northern multinational divisions. While the military forces of Great Britain and Poland took the lead in this area and provided command and control for MND-SE and

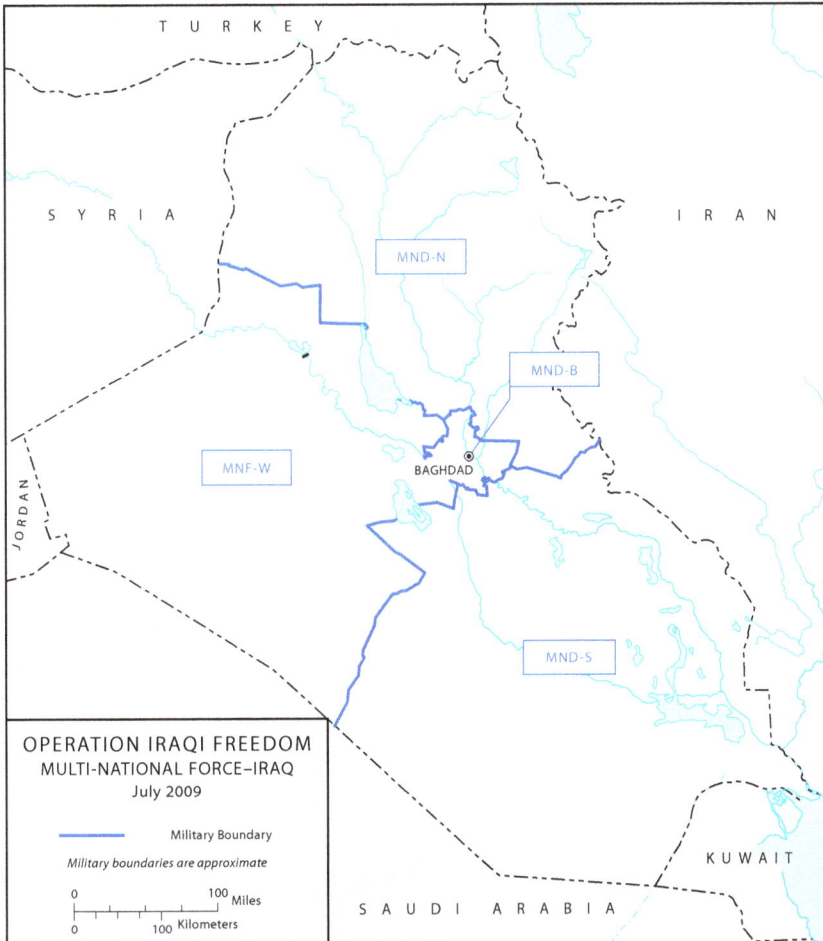

OPERATION IRAQI FREEDOM
MULTI-NATIONAL FORCE–IRAQ
July 2009

Military Boundary

Military boundaries are approximate

0   100 Miles
0   100 Kilometers

Map 6

MND-CS, many other partners filled in with vital security forces. Indeed, the number of allied nations that directly contributed soldiers to Iraq was consistent with the level of coalition partners in the Vietnam and first Gulf Wars.[7] (*See Table 1.*)

---

[7] Thirty-seven nations contributed troops to the Vietnam conflict, while thirty-two allied partners took part in the first Gulf War.

Another issue is also apparent. From 2005 on, critics of the war focused on the withdrawal of various national contingents, claiming that the Coalition of the Willing was crumbling and that America's allies had lost faith in the Iraq mission. The departure of the Spanish and Italian contingents did, to a degree, reflect the unpopularity of the operation among their own national constituents; however, this view ignores a number of issues. Most allied partners deployed forces to Iraq with either an explicit mandate on the length of that deployment or with a specific mission to accomplish before departing. Thailand, for example, deployed forces to Iraq in September 2003 with the understanding that its soldiers would remain in Iraq for one year, in two six-month deployments. Accordingly, Thai forces withdrew in September 2004, after satisfying the commitment. In contrast, the Netherlands deployed military forces with a specific mission. First arriving in August 2003, Dutch troops served in Al Muthanna Province with the explicit task of reconstructing the province and returning it to sovereign Iraqi control. Only in March 2005, after nineteen months of operations in the province and after Al Muthanna was judged "pacified" and under Iraqi authority, did the Dutch forces withdraw.

There were other reasons allied partners withdrew. Nicaragua, for example, ran out of funds to continue its overseas deployment. The Philippines pulled its forces out after insurgents threatened to kill a Filipino hostage. The military contingent from the Republic of Georgia raced back to its home country after violence erupted with Russia. After Spain withdrew following the Madrid bombings in March 2004, two nations in the Spanish-led Plus Ultra Brigade also left Iraq. Equally significant, many nations that withdrew their forces at the end of their national mandates in Iraq redistributed those forces to other missions. For example, a number of nations continued providing troops to the NATO Training Mission–Iraq or United Nations Assistance Mission for Iraq. Others removed troops from Iraq only to increase their participation in operations in Afghanistan. Examples of this redistribution of forces to Afghanistan include, but are not limited to, Albania, Armenia, Azerbaijan, Bosnia-Herzegovina, Bulgaria, Estonia, Georgia, Hungary, Latvia, Poland, Republic of Korea, Romania, Mongolia, and Slovakia.

The increase of capable Iraqi Security Forces also affected coalition troop levels in Iraq. After the Coalition Provisional

Authority officially disbanded the Iraqi Army in May 2003, Iraq lacked any military capability, forcing the United States and its allied partners to satisfy even the most basic security missions. But, as more Iraqi Security Forces were created and sufficiently trained and equipped, the number of coalition forces needed for security missions fell dramatically. This enabled the latter to decrease their own force levels, especially in the south where most of the allied forces had been positioned.

Finally, the Iraqi government ruled against retaining allied forces in Iraq past 31 July 2009. As late as August 2008, forces from twenty-one allied nations remained, with many pledging to stay in Iraq as long as the U.S. military commitment lasted. Thus, rather than crumbling, the allied coalition remained remarkably vibrant. In the end it would also provide a firm foundation for greater military support for the related and somewhat similar tasks in Afghanistan as that conflict began to receive increased attention.

Appendix

Force Contributions by Nations

The remainder of this monograph focuses on the individual contributions of each of the thirty-seven allied partners that maintained troops in Iraq at some time between March 2003 and July 2009. The separate section on each coalition ally presents basic information about the deployed military forces and their general experiences in Operation IRAQI FREEDOM. In some instances, the information available is an approximation at best. Sometimes the internal concerns of allied governments limited the amount of information obtainable; in other cases, such information was not yet in final form. Nevertheless, these sections provide valuable summaries of the important contributions made to the mission in Iraq by America's international partners.

ALBANIA

Ground troops deployed (cumulative): 1,320
Peak deployment: 240
Deployment dates: 6 April 2003–17 December 2008
Unit designation: Not applicable
Order of battle: MND-N
Lead: United States
Primary deployment location(s): Mosul Airfield
Casualties (dead): 0
Casualties (wounded): 5

Force overview: On 6 April 2003, Albania dispatched 70 soldiers to assist with peacekeeping in Iraq prior to the fall of Baghdad. The contingent increased to 120 in 2005, a level that Albania maintained into 2008. In September 2008, Albania increased its deployment to 240. After the United Nations' mandate expired, Albania withdrew all of its forces on 18 December 2008. For most of its six-month deployment rotations, Albania contributed two special forces teams and one infantry company.

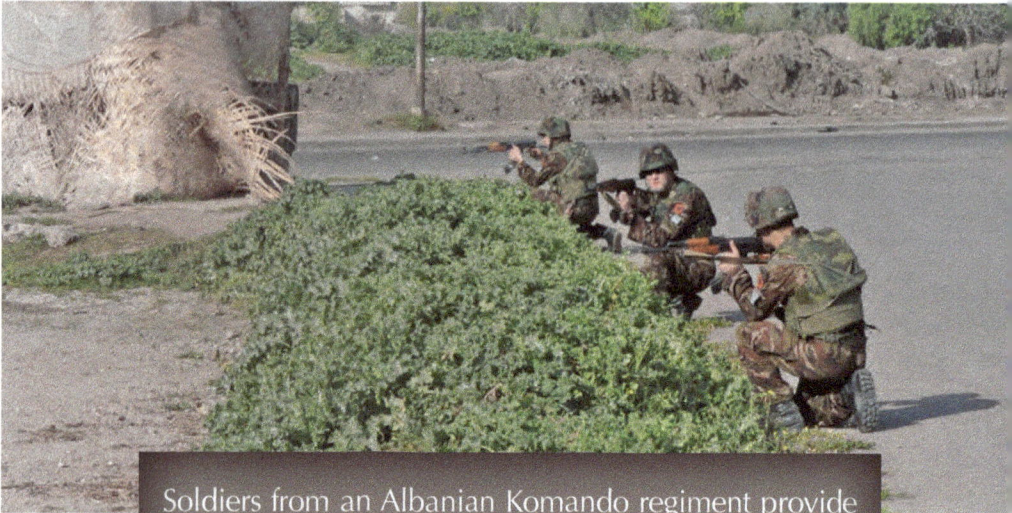

Soldiers from an Albanian Komando regiment provide security along a roadside in Mosul, 29 March 2005.

Operations: The Albanian force's primary role was security near Mosul Airfield, where it sought to minimize the threat to inbound and outbound aircraft. It was also responsible for escorting convoys, guarding checkpoints, maintaining public order, and conducting security patrols.

The Albanian contingent took part in many operations during its deployment. On 25 April 2003, for example, the initial force provided security for four convoys of the 101st Airborne Division headquarters, containing 318 vehicles. The forces also conducted numerous urban patrols to locate weapons and ammunition, apprehended persons possessing weapons caches, maintained order, and secured fuel stations and other critical infrastructure such as railroads. While Albania committed to maintaining troops in Iraq for as long as the United States maintained a presence there, it withdrew in December 2008 at the expiration of the UN mandate.

Armenia

Ground troops deployed (cumulative): 372
Peak deployment: 50
Deployment dates: January 2005–October 2008
Unit designation: Not applicable
Order of battle: MND-CS
Lead: Poland
Primary deployment location(s): Camp Delta, Al Kut
Casualties (dead): 0
Casualties (wounded): Unknown

Force overview: The Armenian Peacekeeping and Humanitarian Mission in Iraq began in January 2005 when the first contingent arrived in MND-CS and was stationed at Camp Delta, near Al Kut, Iraq. The Armenian contingent consisted of a transportation platoon, an engineer team, and a medical team. The transportation platoon provided twenty armored trucks and drivers for convoy missions within MND-CS and from Camp Delta to Kuwait. The engineer team performed road reconnaissance, storage and destruction of explosive materials, and road clearance of

Coalition forces at Camp Victory commemorate the end of mission for the Armenian contingent, 6 October 2008.

unexploded ordnance, as well as serving as part of the Camp Delta Quick Reaction Force with the Salvadoran Battalion. Additionally, the Armenian medical team worked in the Polish 2d Level Field Hospital at Camp Echo and provided medical assistance to the local Iraqis in Ad Diwaniyah and Al Kut.

Operations: The primary roles of the Armenian contingent were peace and stability. In addition, Armenia's interests in Iraq included the welfare and security of the Armenian community in that country, estimated at thirty thousand Iraqis, who resided primarily in Baghdad, Al Basrah, and Mosul. All necessary information was communicated to the U.S. forces in the region to ensure the safety of the Armenian population in Iraq.

Armenia provided extensive assistance between 2005 and 2008. In 2007, for example, the contingent performed 211 surgeries and provided ambulatory care for 680 civilians and 420 military personnel, while running 45 convoys and removing about 9,000 rounds of unexploded ordnance. Armenia withdrew its troops in October 2008.

AUSTRALIA

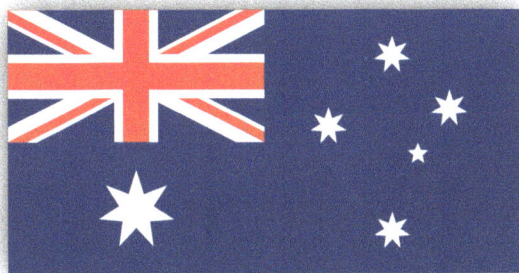

Ground troops deployed (cumulative): 2,400
Peak deployment: 515
Deployment dates: March–May 2003; June 2005–July 2009
Unit designation: Special Forces Task Group; Al Muthanna Task
    Group; Overwatch Battle Group–West
Order of battle: MND-SE
Lead: United Kingdom
Primary deployment location(s): Al Muthanna; Tallil
Casualties (dead): 2
Casualties (wounded): 27

Force overview: Australia contributed forces to several phases of
Operation IRAQI FREEDOM. Operation FALCONER was their first
deployment and participation in major combat operations. The
ground force contribution to this operation was a Special Forces
Task Group composed of five hundred special forces personnel
from the 1st Squadron Group; Australian Special Air Service
Regiment, 4th Battalion; Royal Australian Regiment (Commando),
D Troop; and support units.

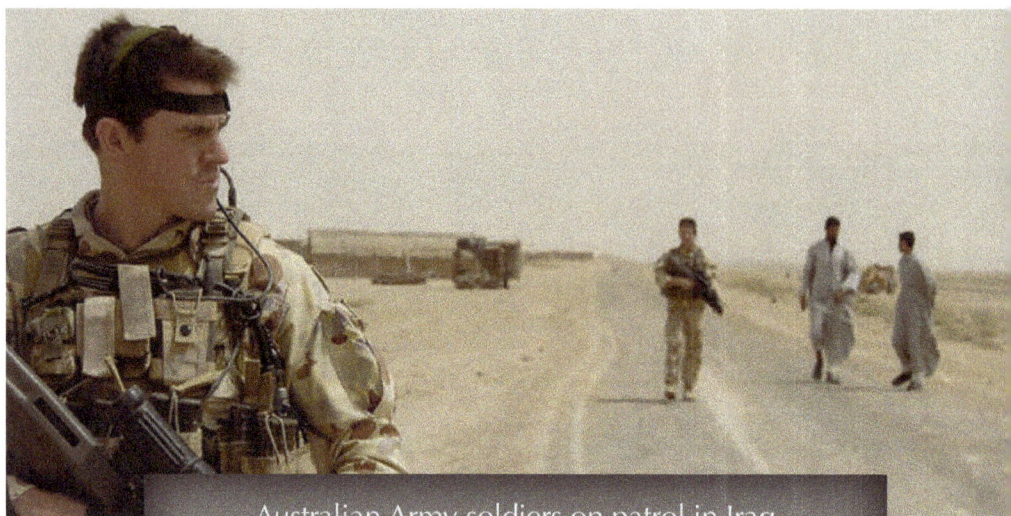
Australian Army soldiers on patrol in Iraq

Most Australian combat forces withdrew at the conclusion of combat operations in May 2003. However, in April 2005, Australia deployed the Al Muthanna Task Group (AMTG) that initially provided security for the Japan Ground Self-Defense Force reconstruction team. This highly mobile team consisted of a cavalry squadron, infantry company, training team, and support units, totaling 450 personnel, 40 Australian light armored vehicles, and 10 Bushmaster vehicles.

In July 2006, after the withdrawal of Japanese reconstruction troops, the AMTG relocated to Dhi Qar Province and was redesignated Overwatch Battle Group–West. OBG-W was located at Tallil Air Base to support Iraqi Security Forces. The battle group was assigned to MND-SE and grew to over five hundred personnel at the end of 2006. The force withdrew from Iraq in June 2008.

Operations: The primary role of Australia's Special Forces Task Group during Operation FALCONER (the initial invasion in March 2003) was to secure an area of western Iraq to prevent possible Scud missile launches against Kuwait, Saudi Arabia, or Israel. The Australian contingent also patrolled highways in western Iraq to block the escape of members of the Hussein regime and to prevent

40

foreign enemy fighters from entering Iraq. On 11 April 2003, the Australian special forces reconcentrated their units into a strike force and captured the Al Asad Air Base, where they seized some fifty MiG jets and millions of pounds of explosives. The forces remained at the base until the end of combat operations in May 2003, when they withdrew or transferred to Baghdad to protect Australian diplomats.

Since Australia did not immediately contribute ground forces to the postwar reconstruction of Iraq, its contribution from May 2003 to June 2005 was limited to a few specialists at the coalition headquarters in Baghdad and to units searching for weapons of mass destruction. Australia also supplied an army training force and a small medical detachment during this time. When Australia redeployed ground troops to Iraq in June 2005 as the Al Muthanna Task Group, they took on two new missions. First, the AMTG provided security for the Japanese Iraq Reconstruction and Support Group after the withdrawal of the Netherlands (which previously provided that support to the Japanese). The AMTG also assisted in the training of local Iraqi Army units.

The Australian mission changed again in July 2006 when the Al Muthanna Task Group moved to Tallil Air Base in July 2006 and was designated Overwatch Battle Group–West. The primary mission of OBG-W was to provide security support if requested by the Iraqi government or Multi-National Force–Iraq. It was never called upon to act in this role; however, the battle group did provide basic training to Iraqi Army personnel at the Basic Training Centre in Tallil. Australia's ground mission in Iraq came to an end in June 2008 when OBG-W withdrew.

Through June 2009, Australia continued to deploy roughly three hundred personnel in Iraq, including a 110-member security detachment to protect the Australian Embassy in Baghdad, 95 liaison officers, a few members of the Coalition Counter IED Task Force, and some support personnel. All Australian forces, except security detachment personnel, withdrew from Iraq on 28 July 2009.

Other military contributions: At the time of this writing, the Royal Australian Navy still deploys frigates to patrol the Persian Gulf. Also, the Royal Australian Air Force has maritime patrol aircraft and C–130 Hercules transport aircraft available to support Operation IRAQI FREEDOM.

AZERBAIJAN

Ground troops deployed (cumulative): 1,100
Peak deployment: 175
Deployment dates: May 2003–December 2008
Unit designation: Not applicable
Order of battle: MNF-W
Lead: United States
Primary deployment location(s): Hadithah
Casualties (dead): 1
Casualties (wounded): Unknown

Force overview: Azerbaijan supplied an infantry company of 150 soldiers in support of Operation IRAQI FREEDOM in May 2003. Beginning with the second rotation, each Azerbaijani contingent completed six months of training in Azerbaijan, focusing on security, marksmanship, crowd control, and communications (including English-language study) prior to its deployment. Azerbaijani soldiers served either six months or one year in Iraq, depending on their enlistment contracts. On 14 November 2008, Azerbaijan's parliament voted to withdraw its force from Iraq, with its troops leaving in December 2008.

An Azerbaijani soldier closes the gate at entry control point 1, Hadithah dam, 17 November 2008.

Operations: Since 15 August 2003, an Azerbaijani infantry company, which was embedded with a U.S. Marine detachment, protected the Hadithah hydroelectric dam, one of Baghdad's main sources of electricity. Located on the western side of the country, it is the largest dam on the Euphrates River. The ten-story structure, designed by Soviet engineers in the 1980s, generates electrical power for the majority of Iraqis living in Al Anbar Province and as much as one-third of all Iraq's power. About 300 Iraqi citizens work as full-time employees at the dam to maintain it. The dam is an essential element in Al Anbar's economic development.

The Azerbaijani infantry company provided perimeter security and force protection for the dam in order to facilitate an uninterrupted power supply and irrigation for Al Anbar Province.

The primary tasks of the Azerbaijan contingent included providing perimeter security for Camp Hadithah, conducting searches and processing of all Iraqi dam workers, escorting all Iraqi workers, and patrolling the tunnel systems within the Hadithah dam. From 2003 on, Azerbaijan deployed over eleven hundred troops to Iraq in support of Operation IRAQI FREEDOM. By providing security to one of Iraq's key infrastructural components, Azerbaijani soldiers freed U.S. forces to conduct security and stability operations in the Hit-Hadithah corridor, as well as the rest of Al Anbar Province.

Other military contributions: Strategically bordering the countries of Iran, Turkey, Russia, Armenia, and Georgia, Azerbaijan has continued to allow the United States use of its air bases. In addition to the contingent in Hadithah, Azerbaijan also had one liaison officer stationed at Camp Victory in Baghdad.

Bosnia-Herzegovina

Ground troops deployed (cumulative): 295
Peak deployment: 85
Deployment dates: June 2005–November 2008
Unit designation: Not applicable
Order of battle: MND-CS; MND-B
Lead: Poland; United States
Primary deployment location(s): 36 soldiers involved in ordnance disposal in Ad Diwaniyah; 49 soldiers to help guard Camp Victory in Baghdad
Casualties (dead): 0
Casualties (wounded): 0

Force overview: Bosnia-Herzegovina sent an explosive ordnance disposal group of thirty-six men to Iraq on 10 June 2005. In August 2008, Bosnia-Herzegovina deployed an additional forty-nine soldiers to guard Camp Victory in Baghdad, increasing its contribution to a total of eighty-five.

Operations: Bosnia-Herzegovina's primary mission in Iraq was to provide ordnance disposal. Its mandate grew during the

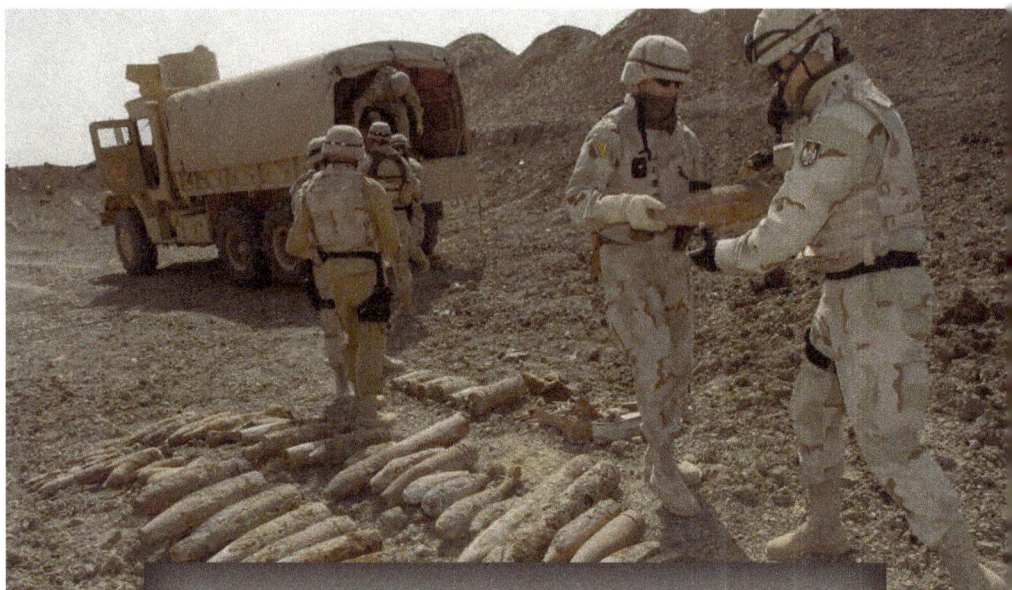

Explosive ordnance disposal soldiers of the Army Forces of Bosnia-Herzegovina prepare the day's find for detonation after a sweep near Tallil, 21 March 2006.

three years of its deployment to include force protection for its EOD personnel and to provide security at Camp Victory. Bosnia-Herzegovina's initial EOD platoon deployed to Al Fallujah in June, where it served alongside U.S. marines. In December 2005, the contingent relocated to Tallil Air Base in An Nasiriyah to clear all conventional unexploded munitions from that site. The platoon completed its mission in An Nasiriyah in November 2006 and was reassigned to MND-CS and Camp Echo in Ad Diwaniyah to clear other locations of unexploded munitions. While the platoon was stationed at Ad Diwaniyah, the Bosnian government deployed an additional infantry platoon to Iraq in August 2008 to provide checkpoint security at Camp Victory. When sufficient numbers of Iraqi Security Force EOD teams became available, Bosnia-Herzegovina withdrew its forces in December 2008.

BULGARIA

Ground troops deployed (cumulative): 1,110
Peak deployment: 496
Deployment dates: August 2003–December 2005; March 2006–December 2008
Unit designation: Not applicable
Order of battle: MND-CS
Lead: Poland (after Polish withdrawal, United States)
Primary deployment location(s): Ad Diwaniyah; Camp Grizzly, Ashraf
Casualties (dead): 13
Casualties (wounded): 64

Force overview: Bulgaria initially deployed 400 soldiers to Iraq in August 2003. They were withdrawn in December 2005, but the national leadership sent another 164 personnel in March 2006. The new mission's mandate was set to expire after a year, but it was extended to December 2008. Prior to deployment and in preparation for the tasks ahead, all Bulgarian contingents undertook two months of special training and a further two weeks of special preparation under the tutelage of U.S. Army instructors in military police and prison guard duties.

By June 2008, there were 154 Bulgarian personnel in Iraq. One infantry company was embedded with the U.S. Army at Camp Grizzly in Ashraf, Iraq, providing force protection to the camp, as well as to Camp Ashraf. All Bulgarian forces redeployed home in December 2008.

Operations: Bulgarian soldiers provided security for Camp Grizzly, a refugee camp on the border with Iran, and guards at Camp Ashraf, a temporary facility for interviewing and processing detainees. In addition, Bulgarian forces conducted several convoy-escort missions and patrols. The soldiers patrolled the Tampa supply route, organized convoys, escorted people and cargo, and supported the 1st Battalion, 1st Division, of the Iraqi Army.

Bulgaria's force suffered thirteen deaths and sixty-four personnel wounded during its five-year deployment, including five deaths on 27 and 28 December 2003 in two separate car bomb attacks.

Other military contributions: As of this writing, the Bulgarian Army still provides four officers to the NATO Training Mission–Iraq.

CZECH REPUBLIC

Ground troops deployed (cumulative): 2,000
Peak deployment: 357
Deployment dates: April 2003–December 2008
Unit designation: Not applicable
Order of battle: MND-SE
Lead: United Kingdom
Primary deployment location(s): Az Zubayr; Al Basrah; Baghdad
Casualties (dead): 1
Casualties (wounded): Not applicable

Force overview: The Czech Republic first deployed troops to Iraq in April 2003. The original Czech contingent consisted of three hundred troops and three civilians. The group established and operated the 7th Field Hospital, just outside of Al Basrah. In addition to the hospital personnel deployed in Iraq, a unit from the 1st Czech-Slovak Battalion of Radiation, Chemical, and Biological Protection was positioned in Kuwait at the start of combat operations in case the Iraqi military unleashed such weapons. The 7th Field Hospital was later replaced by a contingent of eighty military police assigned

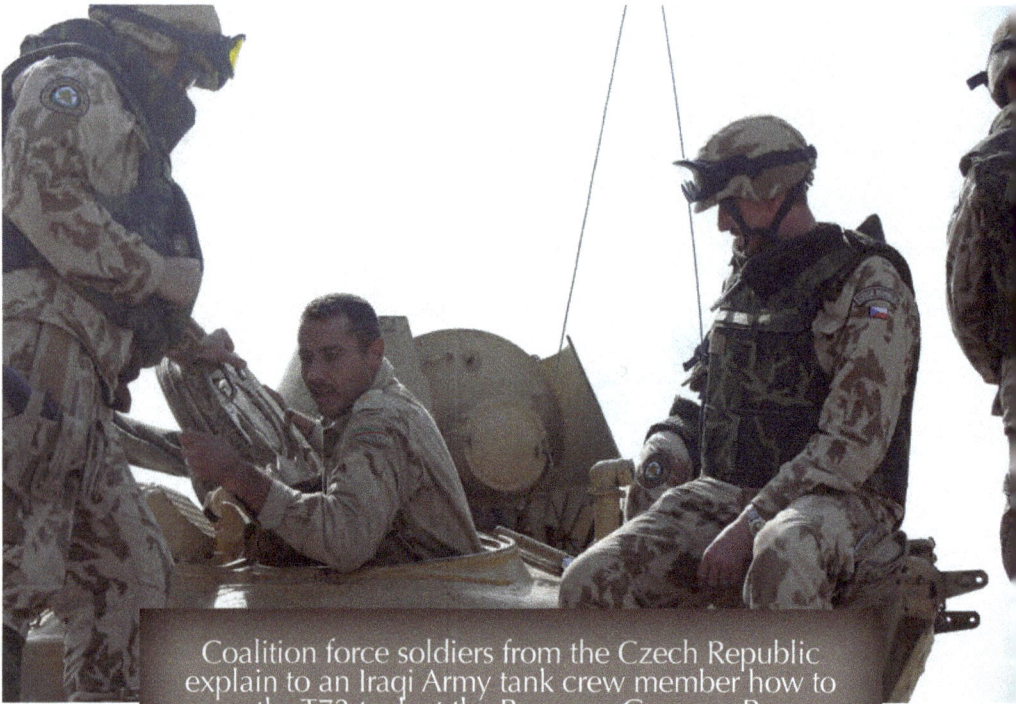

Coalition force soldiers from the Czech Republic explain to an Iraqi Army tank crew member how to zero the T72 tank at the Besmaya Gunnery Range, 27 October 2008.

to the Shaibah Base, where they operated under the command of the U.K. in what would become MND-SE. Beginning in January 2007, the Czech contingent began force protection operations at the Al Basrah Air Station at vehicle checkpoints. By December 2007, the contingent was reduced to 100 and then reduced again to 17 during the summer of 2008. All remaining Czech troops were withdrawn from Iraq by the end of 2008.

Operations: During the initial Czech deployment at the 7th Field Hospital, Czech personnel struggled to construct the hospital and ensure sufficient quantities of fresh water in the marshy land around Al Basrah. Once the hospital was operational, they treated large numbers of coalition and Iraqi casualties.

The second Czech rotation, composed of military police personnel, worked at the Shaibah Base to construct checkpoints, supervise local police, help train Iraqi local constabulary at the national academy at Az Zubayr, and contribute security support to MND-SE. The military police continued that role until December 2006 when the police academy shut down.

After the closure of the police academy at Az Zubayr, follow-on Czech contingents provided traffic control into and out of the Shaibah Base, force protection of that base, and mentoring to the Iraqi police staff in Al Basrah. The contingent was embedded with the British Royal Air Force–Force Protection Wing at Contingency Operating Base Al Basrah.

During the nearly six years of their deployment, the Czech medics and military police sought to ensure Iraqi police were ready to handle their country's security. The Czech contributions to MND-SE were significant.

Other military contributions: The Czech Republic provided five personnel in support of the NATO Training Mission–Iraq.

DENMARK

Ground troops deployed (cumulative): 5,500
Peak deployment: 545
Deployment dates: April 2003–December 2007
Unit designation: Dancon/Irak mission
Order of battle: MND-SE
Lead: United Kingdom
Primary deployment location(s): Camp Eden; Camp Danevang, Al Basrah
Casualties (dead): 7
Casualties (wounded): Unknown

Force overview: The first Danish troop deployment consisted of 380 personnel, including medical, military police, and infantry forces. Another 42 soldiers arrived in August 2003, bringing the total force to 422. By February 2005, the contingent consisted of 545 soldiers and remained at that level until March 2007, when it fell to 460. In mid-2005, Denmark also sent a detachment of four Fennec helicopters to Iraq for reconnaissance and transportation

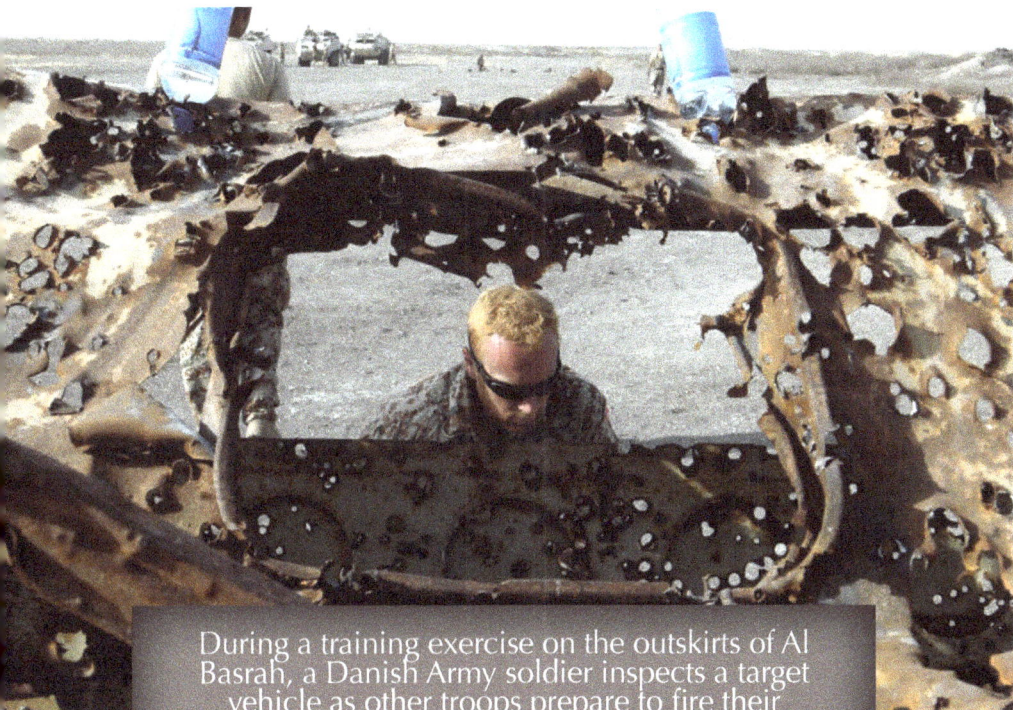

During a training exercise on the outskirts of Al Basrah, a Danish Army soldier inspects a target vehicle as other troops prepare to fire their weapons, 5 August 2005.

as part of the British Joint Helicopter Detachment that performed nightly surveillance missions. In June 2007, fifty-five Danish air force personnel replaced the ground contingent.

Operations: The Danes were responsible for a variety of missions: conducting transport and mission support, searching for biological weapons, monitoring prisoners at the Camp Eden detention facility, and completing civilian reconstruction projects. Denmark's initial deployment was to Camp Eden, where its forces remained for fourteen months before moving to Camp Danevang, near Al Basrah. Since 2003, Danish forces completed more than 680 reconstruction projects with an overall budget of $3.2 million.

On 31 July 2007, Denmark transferred responsibilities for its missions to U.K. forces in the region. In August of that year,

Denmark withdrew the majority of its soldiers but kept its four Fennec helicopters and support personnel at Al Basrah Air Station until December 2007.

Other military contributions: Denmark also provided frigates and submarines to patrol the Persian Gulf; deployed thirty-five soldiers to serve as guards for the United Nations in Baghdad; and beginning in 2005, contributed ten instructors and seven guards to assist the NATO Training Mission–Iraq.

DOMINICAN REPUBLIC

Ground troops deployed (cumulative): 600
Peak deployment: 302
Deployment dates: April 2003–May 2004
Unit designation: Part of Multi-National Brigade Plus Ultra
Order of battle: MND-CS
Lead: Poland (under direct Spanish control)
Primary deployment location(s): An Najaf
Casualties (dead): 0
Casualties (wounded): 0

Force overview: The Dominican Republic deployed 302 soldiers from April 2003 until May 2004. Its forces joined those of Spain, El Salvador, Honduras, and Nicaragua in the Plus Ultra Brigade, or Brigada Hispanoamericana, which consisted of some 2,500 troops at its peak. The Dominican troops manned Base Santo Domingo in An Najaf.

Operations: The Dominican Republic was one of the first nations to send troops to Iraq after the close of major combat operations.

Dominican soldiers provided humanitarian assistance and helped to rebuild city infrastructure in An Najaf. The troops faced frequent mortar and rocket attacks against their base during the course of the deployment but suffered no casualties. In 2004, the Dominican government withdrew its forces about one month after Spain and Honduras, citing domestic opposition to its military involvement.

EL SALVADOR

Ground troops deployed (cumulative): 5,800
Peak deployment: 380
Deployment dates: August 2003–22 January 2009
Unit designation: Cuscatlán Battalion; part of Multi-National
    Brigade Plus Ultra from August 2003 to April 2004
Order of battle: MND-CS
Lead: Poland (under direct Spanish control from August 2003 to
    April 2004)
Primary deployment location(s): Base El Salvador, An Najaf
Casualties (dead): 5
Casualties (wounded): 20

Force overview: El Salvador sent a light infantry battalion
(the Cuscatlán Battalion) to Iraq in August 2003. The first
contingent was composed of 360 soldiers and joined the Plus
Ultra Brigade.

In May 2004, after Spain and the other nations of the Plus Ultra
Brigade withdrew from Iraq, El Salvador increased its contribution

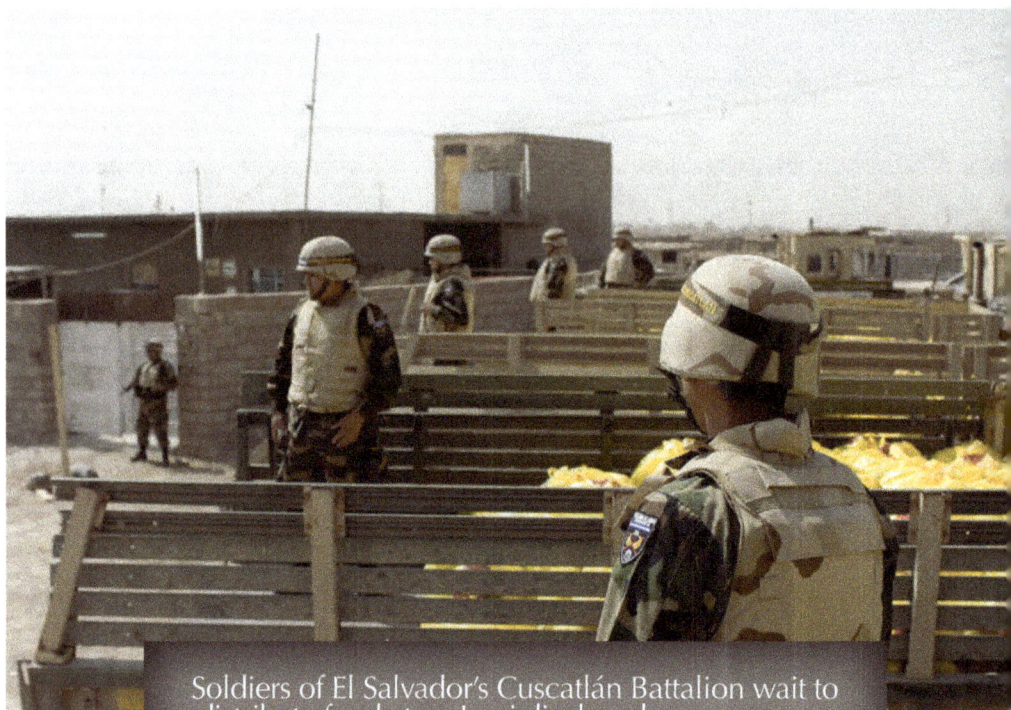

Soldiers of El Salvador's Cuscatlán Battalion wait to distribute food at an Iraqi displaced persons camp near Al Kut, 14 August 2008.

to 380 troops and maintained that strength until 2008. It was the only Central or South American country that sustained its forces in Iraq throughout the entire coalition involvement. Salvadoran personnel withdrew from Iraq on 22 January 2009.

Operations: Most of the duties of the Salvadoran force focused on humanitarian missions and the rebuilding of the Iraqi infrastructure. For example, during its four months of deployment in 2003, it supervised 130 humanitarian missions, including the distribution of food and clothing and the reconstruction of schools, roads, medical facilities, and water-treatment plants, while also leading security patrols and escorting convoys. After defending one such convoy that included U.S. and coalition members caught in an ambush in 2004 near An Najaf, six Salvadoran soldiers were awarded the U.S. Bronze Star by Secretary of Defense Donald H. Rumsfeld on 12 November of that year.

El Salvador's President Antonio Saca explained his nation's participation in Operation IRAQI FREEDOM in a speech, noting that "during the 1980s El Salvador fought a prolonged war against terrorist organizations sponsored by the now-defunct Soviet Union through its proxy, Cuba . . . [and] prevailed . . . but that achievement could not have been accomplished without the support of the international community and mainly of the United States. For us, participating in Iraq is more than just payback. Being in Iraq is a way of telling the Iraqi people, 'It is possible to overcome; it is possible to rebuild your country even from the ashes and to procure a future for your children and generations to come.'"

ESTONIA

Ground troops deployed (cumulative): 240
Peak deployment: 40
Deployment dates: 20 June 2003–7 February 2009
Unit designation: Estonian Light Infantry Platoon (the "Stone Platoon")
Order of battle: MND-B
Lead: United States
Primary deployment location(s): Camp Cook, At Taji
Casualties (dead): 2
Casualties (wounded): 20

Force overview: The initial Estonian unit that deployed to Iraq, which departed Tallinn on 20 June 2003 aboard a U.S. Air Force C–17, consisted of an infantry platoon equipped with small arms manufactured by Israel, Germany, Sweden, and the United States, along with three unarmored Unimog trucks. The platoon of thirty-four men arrived at the At Taji Military Training Base just north of Baghdad, where it was tasked with patrolling, conducting cordon-

and-search operations, and providing quick-reaction forces under the command of MND-B. The final Estonian forces withdrew from Iraq on 7 February 2009.

Operations: The Estonian parliament issued a mandate for its participation in international peace-support operations in Iraq in May 2003. Under the aegis of this mandate, the Estonian soldiers were authorized to conduct operations against the former regime extremists and foreign terrorists and assist in the organizing, training, and supplying of the Iraqi armed forces. As conceived in 2003, the mission was for peace support but transitioned into counterinsurgency. The Estonian government continued its support of the Coalition with no significant change until the start of 2009.

The Estonian contingent took part in regular operations in MND-B's area of responsibility, securing its sector and acting as a quick-reaction force. The unit earned the nickname the "Stone Platoon" during operations in Al Fallujah in November 2004 as the Estonian soldiers moved through the besieged city driving their unarmored Unimogs, while most coalition forces had up-armored vehicles; however, Estonian defense-force technicians soon outfitted their Unimog trucks with armored machine-gun turrets and bulletproof glass windows.

The Estonian contingent took part in numerous operations. During Operation BIG DIG (January 2005) in the Latifiyah area of northern Babil Province, just south of Baghdad, members of the Stone Platoon uncovered a cache of seven 1,000-kilogram warheads. In 2007 alone, the Stone Platoon conducted some 15 cordon-and-search operations, over 300 combat patrols and 350 external patrols, as well as detained 28 insurgents and located 15 caches of weapons, including hundreds of rockets and other explosives.

Other military contributions: Three Estonian staff officers served with NATO Training Mission–Iraq to help train Iraqi Security Forces.

GEORGIA

Ground troops deployed (cumulative): 10,000
Peak deployment: 1,850
Deployment dates: 3 August 2003–10 August 2008
Unit designation: 3d Infantry Brigade
Order of battle: MND-B; MND-C
Lead: United States
Primary deployment location(s): Camp Delta, Al Kut; FOB Warhorse; FOB Caldwell; FOB Gabe
Casualties (dead): 5
Casualties (wounded): 27

Force overview: The first Georgian contingent, composed of seventy military medical personnel and special forces, arrived in Iraq on 3 August 2003. The special forces detachment deployed to Bayji, 130 miles north of Baghdad, while the medical team moved to Balad, a primarily Shi'ite town in the Sunni Triangle. In 2004, the Georgian presence increased to three hundred personnel with the arrival of the 16th Mountain Battalion and the 12th Commando Brigade.

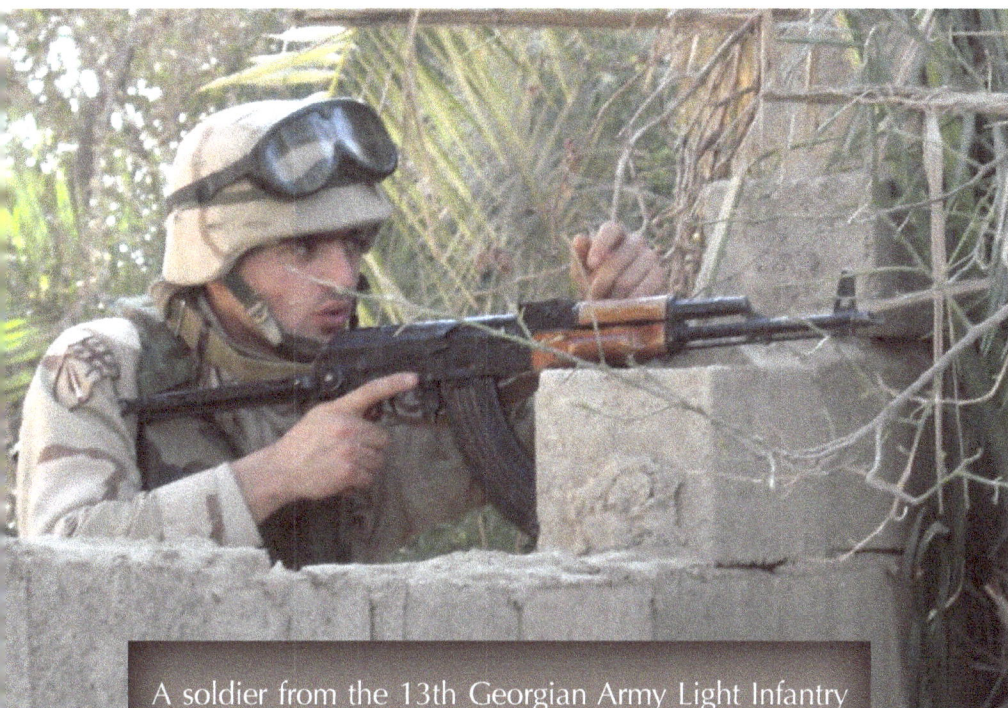

A soldier from the 13th Georgian Army Light Infantry Battalion provides security from behind a wall during a patrol through Al Lej, 7 March 2008.

Both deployed to Ba'qubah, northeast of Baghdad and just outside the Sunni Triangle, an area of heavy insurgent activity. In 2005, the number of Georgian troops stationed in Iraq again increased to over 850 and, between 2005 and 2007, the Georgian forces split their operations between Ba'qubah and the International Zone (IZ) in Baghdad. On 15 July 2007, the Georgian military presence increased to over 2000, with 150 troops providing security for the UN compound in Baghdad and the remaining 1,850 deployed to Al Kut, southeast of Baghdad, at Camp Delta. In January 2008, the Georgian 3d Infantry Brigade was replaced by the 1st Brigade.

Operations: From 2003 to 2007, the Georgian mission focused on key infrastructure security, checkpoints, base protection, patrols, and security in the International Zone, including the UN compound in Baghdad (see below). The mission changed considerably, however, in 2007 when the 3d Infantry Brigade deployed to Al Kut near the Iranian border. While about one hundred troops remained in the IZ for security, the 3d Brigade's mission was to interdict supplies being smuggled to Shi'ite extremists from Iran. During this time, the Georgians established six patrol bases and began to patrol intensively along the Iraq-Iran border.

Georgian forces withdrew from Iraq on 10 August 2008 in response to the outbreak of fighting with Russia in the breakaway Georgian province of South Ossetia. The Georgians were replaced by Iraqi National Police and the 32d Iraqi Army Brigade. While the Georgian departure was unexpected, the response of Iraqi replacements highlighted the increasing capabilities of the Iraqi Army.

Other military contributions: Georgia furnished several battalions, up to 558 troops per deployment, to provide security for the UN compound in Baghdad. Units that participated included the 13th (Shavnabada) Light Infantry, 21st Light Infantry, 22d Light Infantry, and 33d Light Infantry Battalions. By 2007, the mission was reduced to a single company of 150 soldiers securing the compound.

HONDURAS

Ground troops deployed (cumulative): 736
Peak deployment: 368
Deployment dates: August 2003–April 2004
Unit designation: Part of Multi-National Brigade Plus Ultra
Order of battle: MND-CS
Lead: Poland (under direct Spanish control)
Primary deployment location(s): Base Tegucigalpa, An Najaf
Casualties (dead): 0
Casualties (wounded): 0

Force overview: In August 2003, Honduras deployed 312 soldiers to join the Plus Ultra Brigade, or Brigada Hispanoamericana, which included forces from Spain, El Salvador, Nicaragua, and the Dominican Republic. Honduras increased its troop contribution to 368 soldiers and manned Base Tegucigalpa in An Najaf.

Operations: The Honduran mission focused on reconstruction in and around An Najaf. The Honduran government ordered its troops to withdraw from Iraq in April 2004, citing both a lack of domestic public support and the completion of its reconstruction mission in An Najaf.

HUNGARY

Ground troops deployed (cumulative): 600
Peak deployment: 300
Deployment dates: August 2003–December 2004
Unit designation: Not applicable
Order of battle: MND-CS
Lead: Poland
Primary deployment location(s): Al Hillah
Casualties (dead): 1
Casualties (wounded): Unknown

Force overview: In July 2003, Hungary deployed a military transport battalion composed of 300 troops to Iraq. The battalion joined the Polish multinational division in what became MND-CS in Al Hillah.

Operations: The Hungarian contingent specialized in logistics and the distribution of humanitarian aid to Iraqi civilians. The force lost one soldier in June 2004 when an IED exploded near a Hungarian convoy.

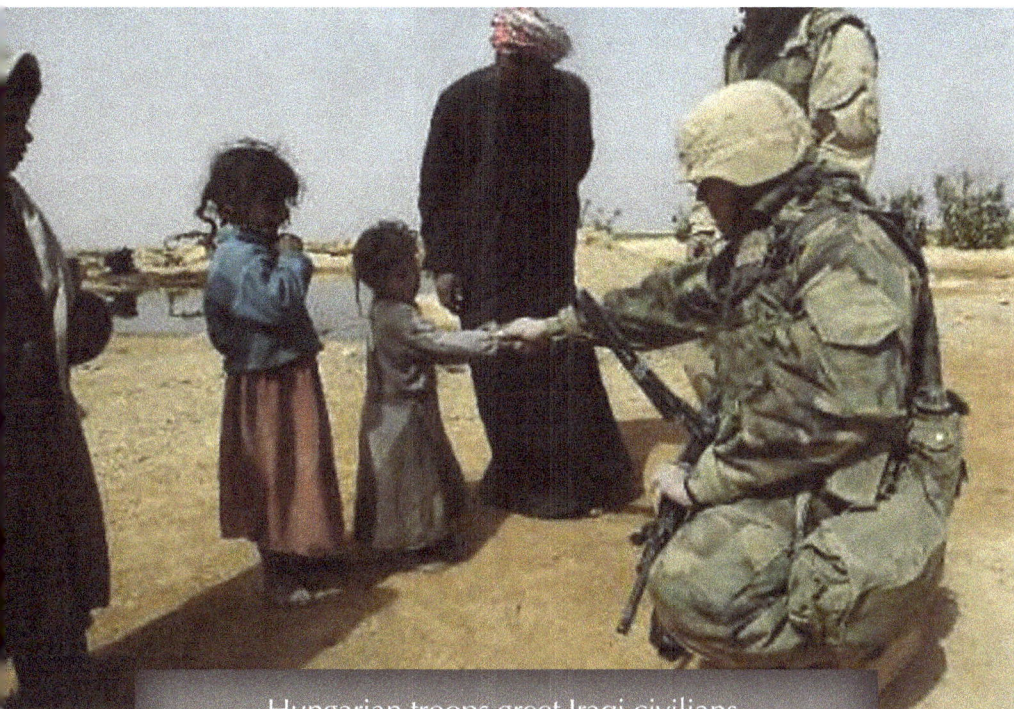
Hungarian troops greet Iraqi civilians.

In December 2004, the Hungarian parliament decided to withdraw its troops rather than wait for the Iraqi elections in January 2005. However, in late 2005, Hungary sent troops to support the NATO Training Mission–Iraq (see below).

Other military contributions: In December 2002, the Hungarian government made available an air base at Taszár to train exiled Iraqi opposition leaders. Hungary provided 150 trainers and security personnel for the NTM-I mission in December 2005, and as of early 2009 had 17 trainers with the mission.

ITALY

Ground troops deployed (cumulative): 7,800
Peak deployment: 2,600
Deployment dates: July 2003–November 2006
Unit designation: Garibaldi Brigade
Order of battle: MND-SE
Lead: United Kingdom
Primary deployment location(s): An Nasiriyah
Casualties (dead): 33
Casualties (wounded): Unknown

Force overview: Italy deployed its first contingent, some 2,400 personnel, in July 2003. The Garibaldi Brigade consisted of mechanized infantry, helicopters, and Carabinieri (the national gendarmerie, or police of Italy). It served primarily in southern Iraq, near An Nasiriyah, as part of the British-led multinational division, which later became MND-SE. The Italian contingent also coordinated missions with a Romanian mechanized infantry battalion, a Romanian military police company, and a Portuguese security company.

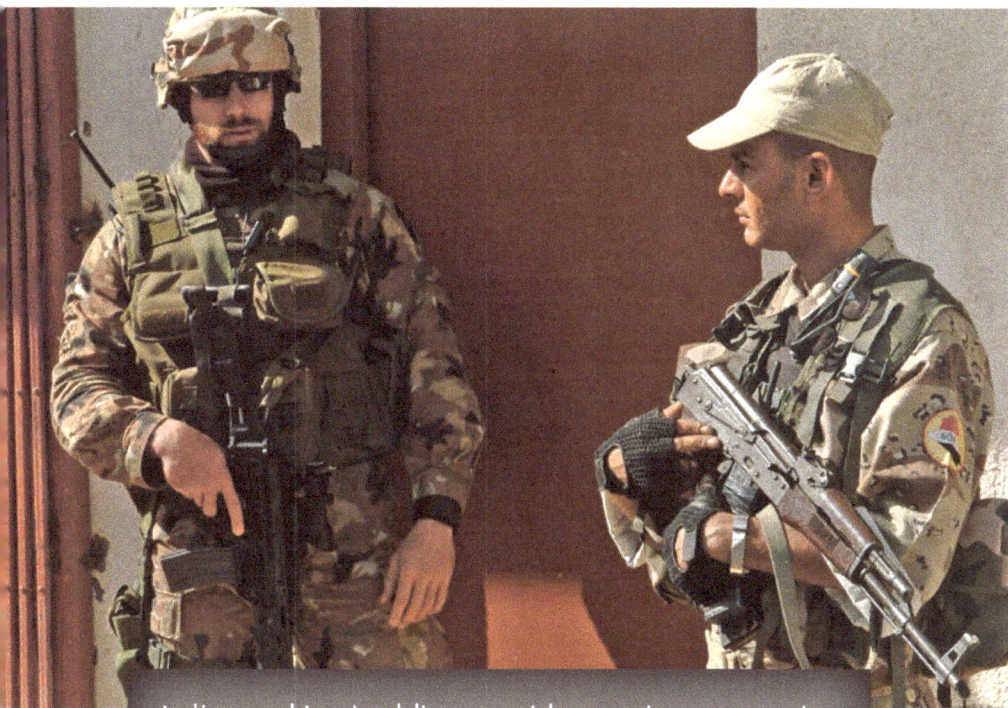

Italian and Iraqi soldiers provide security at a meeting between Italian, Iraqi, and U.S. military leaders at An Nasiriyah, 16 March 2005.

Starting in July 2005, the Italian government began to withdraw soldiers from Iraq at a measured pace, leaving about 1,600 personnel by June 2006, with the rest leaving Iraq by November 2006.

Operations: The Italians provided humanitarian aid to the Iraqi people, training for local police forces, and military police support for MND-SE. During their deployment, Italian forces suffered their most serious losses since World War II on 12 November 2003, when suicide bombers in two explosive-laden vehicles crashed through the main gate at the Italian Carabinieri headquarters in An Nasiriyah. The ensuing blast leveled the building, killing twelve Carabinieri, five Italian soldiers, and two Italian civilians.

Despite the losses, Italian forces conducted many successful operations in 2003 and 2004. For example, in Dhi Qar Province, Italian patrols located large caches of hidden weapons and munitions. During Operation GATHERING SECURITY, they seized a significant number of mortars, mines, and rocket-propelled grenades that were being fabricated into IEDs.

Other military contributions: Italy continues to contribute trainers to the NATO Training Mission–Iraq.

JAPAN

Ground troops deployed (cumulative): 6,100
Peak deployment: 600
Deployment dates: 3 February 2004–25 July 2006
Unit designation: Iraqi Reconstruction Support Group
Order of battle: MND-SE
Lead: United Kingdom
Primary deployment location(s): As Samawah, Al Muthanna Province
Casualties (dead): 0
Casualties (wounded): Unknown

Force overview: The Japanese Ground Self-Defense Force's (JGSDF) Iraqi Reconstruction Support Group deployed in February 2004 to Multi-National Division–Southeast under the direction of the United Kingdom. Japan's force consisted of some six hundred personnel at its peak, primarily engineers and medical staff. While security for the contingent was generally undertaken by Dutch and later Australian troops due to restrictive Japanese rules of engagement, a few soldiers from the Japanese Special

A Japan Air Self-Defense Force soldier salutes before opening the gate to a flight line, 21 June 2006.

Operations Group (Tokushu Sakusen Gun), Western Infantry Army Regiment, and the 1st Airborne Brigade provided security. The Japanese also deployed a Planning and Liaison Unit (16 January 2004–25 July 2006) and a Redeployment Support Unit (26 June–9 September 2006) to assist in the initial troop deployments to Iraq and their return to Japan.

Operations: The role of the JGSDF forces in Iraq was entirely humanitarian. During its seventeen-month deployment, the JGSDF assisted four Iraqi hospitals, including As Samawah General Hospital, in training local doctors in diagnostic methods and provided pharmaceuticals and medical equipment. One of Japan's significant medical accomplishments was to reduce the mortality rate of newborn infants in As Samawah by nearly one-third by the end of its deployment.

The JGSDF's engineers also undertook numerous civil works projects, many of which involved repairing public facilities and infrastructure. For example, during their deployment, the Japanese repaired thirty-six local schools, comprising nearly one-third of those in Al Muthanna Province; improved and paved thirty-one roads; and repaired medical clinics, nursing facilities, and low-income residential housing units in As Samawah. The JGSDF also worked on several cultural sites, including the ruins of Uruk (also known as Erech) and the Olympic Stadium. Additionally, the JGSDF provided vehicles to deliver fresh water to much of the As Samawah camp, constructed a water purification facility near the camp, and rebuilt water purification facilities in the nearby towns of Warka and Ar Rumaythah. In the process, many local citizens were recruited for jobs at the JGSDF As Samawah camp and on many JGSDF-led public works projects. In the end, the Japanese were employing an average of nearly eleven hundred Iraqis per day.

Other military contributions: The Iraq Reconstruction Support Airlift Wing (IRSAW) conducted airlift operations from Ali Al Salem Air Base in Kuwait to Tallil, Iraq, beginning in March 2004. After the JGSDF redeployment in July 2006, the Japan Air Self-Defense Force continued to provide airlift support to the United Nations and multinational forces in Iraq, with flights from Ali Al Salem to Tallil, Baghdad (beginning on 31 July 2006), and Arbil via Baghdad, primarily for UN personnel and goods (beginning on 6 September 2006). As of 31 January 2008, IRSAW had conducted 656 airlifts, transporting 587 tons of goods and materials. The mission expired along with the UN mandate at the end of December 2008.

KAZAKHSTAN

Ground troops deployed (cumulative): 320
Peak deployment: 29
Deployment dates: September 2003–October 2008
Unit designation: Kazbat Peacekeeping Battalion
Order of battle: MND-CS
Lead: Poland
Primary deployment location(s): FOB Kalsu, Al Kut, Wasit Province
Casualties (dead): 1
Casualties (wounded): Unknown

Force overview: Beginning September 2003 and running for more than five years, Kazakhstan kept twenty-seven sappers in Iraq to aid multinational forces in disarming explosive devices. The contingent included EOD experts, engineers, and medical personnel.

Operations: From September 2003 to October 2008, Kazakhstan's military engineers were tasked with searching for and destroying unexploded ordnance in and around the city of Al Kut. In addition, the Kazakh EOD soldiers destroyed or disarmed all ammunition

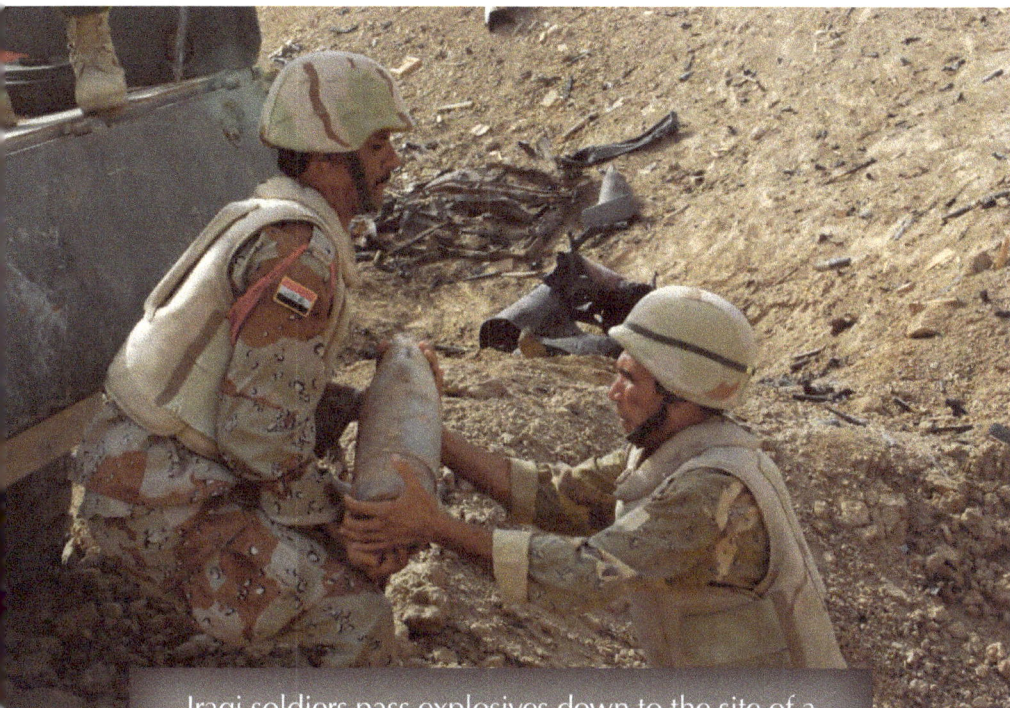

Iraqi soldiers pass explosives down to the site of a controlled detonation by Kazakh soldiers and Iraqi explosive ordnance disposal trainees on Forward Operating Base Delta, 25 September 2008.

captured in regional raids, amounting to nearly 5 million rounds of ordnance. The Kazakh success was not without costs, however. In 2005, a Kazakh EOD engineer and four Ukrainian soldiers were killed when an insurgent-rigged IED exploded.

The Kazakhstan contingent also operated a water purification system; provided medical aid, with assistance from the Polish contingent, to nearly five thousand Iraqis; and trained Iraqi engineers and sappers in EOD techniques and practices. By October 2008, with enough Iraqi personnel trained in ordnance disposal, the Kazakhs handed control over to the Iraqi Security Forces and withdrew along with the Polish forces.

LATVIA

Ground troops deployed (cumulative): 1,150
Peak deployment: 126
Deployment dates: May 2003–November 2008
Unit designation: Not applicable
Order of battle: MND-CS
Lead: Poland
Primary deployment location(s): Camp Echo, Ad Diwaniyah
Casualties (killed): 3
Casualties (wounded): 5

Force overview: Latvia's initial contingent was deployed to Iraq in May 2003. It consisted of a small EOD team and a logistics platoon but quickly grew to 126 soldiers. By the end of June 2007, all Latvian soldiers had departed, leaving only three officers who continued to work in intelligence analysis and operational planning at MND-CS and at Headquarters, Multi-National Corps–Iraq. Those final liaison officers withdrew from Iraq in November 2008, shortly after Poland's withdrawal.

Operations: Latvia's official mission in Iraq was to assist the Iraqi government and the newly established Iraqi forces to take responsibility for their own security. Initially deployed to Kirkuk

U.S. Army and Latvian soldiers provide security for a convoy near Tallil, 6 March 2007.

for one year, the contingent was transferred to Camp Charlie in Al Hillah the following May, then later to Camp Delta in Al Kut. Finally, Latvian forces were stationed at Camp Echo in Ad Diwaniyah, where they remained until June 2007. Aside from their EOD mission, Latvian infantrymen were tasked with providing a quick-reaction force, convoy escorts, combined joint operations, force protection, law-enforcement training, infrastructure development, and local security patrols.

During Latvia's five and a half years in Iraq, its forces suffered three casualties, two of which occurred on the forces' final deployment while patrolling outside of Camp Echo. In June 2007, Latvia withdrew its contingent from Iraq while simultaneously increasing its commitment to Afghanistan.

LITHUANIA

Ground troops deployed (cumulative): 850
Peak deployment: 750
Deployment dates: August 2003–July 2007; October 2007–August 2008
Unit designation: Not applicable
Order of battle: MND-CS; MND-SE; MND-C
Lead: Poland; United Kingdom; United States
Primary deployment location(s): Camp Echo, Al Hillah; Al Basrah
Casualties (dead): 0
Casualties (wounded): 0

Force overview: Starting in 2003, Lithuania provided a total of 850 troops to Iraq in support of Operation IRAQI FREEDOM. Its original contribution, an infantry platoon, was first deployed to Al Hillah in the south and then to Al Basrah in August 2003. The Al Hillah detachment was part of MND-CS, under Polish control and stationed at Camp Echo. A second contingent deployed to MND-SE in Al Basrah, where it served primarily with Danish troops. In October 2007, Lithuania deployed a new contingent of forty

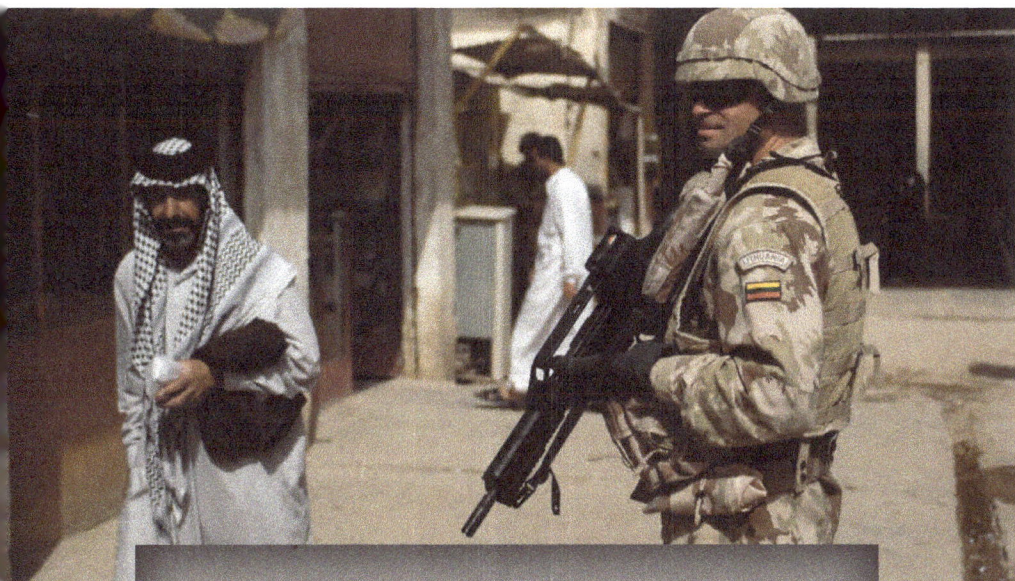

A Lithuanian soldier oversees activity in a market in the town of Ad Dujayl, 20 May 2008.

mounted infantry troops to serve in MND-C under U.S. control, later withdrawing in August 2008.

Operations: Lithuanian missions included combat patrols, checkpoint security, quick-reaction force, support to reconstruction projects, and medical assistance to local Iraqi civilians. In July 2007, Lithuania initially recalled its forces when Denmark withdrew from Iraq, but several months later, in October, it sent a unit of forty soldiers to serve for another six months. In June 2008, eleven soldiers from this contingent left and the remainder stayed for two more months. The Iron Wolf Mechanized Infantry Brigade chose to remain in Iraq rather than withdrawing, so an additional two months were added to the mission. On 1 August 2008, the remaining Lithuanian soldiers returned home.

Other military contributions: Lithuania continues to provide four personnel to the NATO Training Mission–Iraq.

Macedonia

Ground troops deployed (cumulative): 420
Peak deployment: 80
Deployment dates: June 2003–December 2008
Unit designation: Wolves, Scorpions
Order of battle: MND-B
Lead: United States
Primary deployment location(s): At Taji
Casualties (dead): 0
Casualties (wounded): Unknown

Force overview: Since its initial deployment in June 2003, the Macedonian contingent alternated between members of the nation's Ranger Battalion and Special Forces Battalion (known as the Wolves Battalion). Its troops conducted full-spectrum operations, operating independently or with U.S. troops, under MND-B in At Taji, sixty miles north of Baghdad in the heart of the violent Sunni Triangle.

The Macedonian Rangers secured the perimeter around Camp Taji, capturing insurgents, securing communication lines, and confiscating weapons. Meanwhile, a "Wolves" platoon took part in over two hundred missions, including Operations PENINSULA STRIKE and IVY TYPHOON.

Operations: The Macedonian forces conducted a variety of missions while embedded with the U.S. Army's MND-B. These missions included raids, reconnaissance, establishing and manning checkpoints, counter-mortar and counter-rocket patrols, combat patrols, sniper missions and ambushes, quick-reaction force, intelligence gathering, and training of Iraqi Army units. A contingent of Wolves Special Forces personnel participated in Operation PENINSULA STRIKE on 12 June 2003. The Macedonians conducted a sweep that extended from Ad Duluiyah to Tikrit. During the operation, the Wolves arrested three "high-priority individuals" who were on the coalition's list of the "55 most wanted." In Operation IVY TYPHOON, in January 2004, the Wolves platoon carried out raids to locate weapons caches in the area, relying on a combination of local intelligence and reconnaissance. In one instance the Macedonians, supporting the U.S. 5th Engineer Battalion, located and seized a substantial weapons cache and a factory for the construction of IEDs. By July 2007, Macedonian special forces were also participating in air assault operations while continuing their training mission. They trained more than five hundred Iraqi soldiers in 2007 alone.

Other military contributions: Macedonia allowed the United States use of its airspace to fly support and combat missions. In addition, Macedonia provided medical teams to assist in the postwar reconstruction of Iraq.

MOLDOVA

Ground troops deployed (cumulative): 110
Peak deployment: 20
Deployment dates: September 2003–December 2008
Unit designation: Not applicable
Order of battle: MND-N
Lead: United States
Primary deployment location(s): Mosul
Casualties (dead): 0
Casualties (wounded): 0

Force overview: The Moldovan humanitarian mission in Iraq began in September 2003, when its first contingent of twelve personnel, composed of one infantry platoon and an EOD team, was attached to the U.S. 4th Infantry Division's 29th Field Artillery Battalion (Task Force Pacesetter), at Samarra East Airfield. In 2006, the Moldovan contingent relocated to Camp Grizzly, just outside of Ashraf, and in August 2008, Moldova increased its contribution to twenty personnel by sending additional engineers.

Operations: The essential tasks provided by the Moldovan contingent consisted primarily of detecting, identifying, collecting, and disposing of unexploded ordnance and improvised explosive devices. The troops also supplied mine-awareness training to all multinational forces, EOD training to Iraqi Army personnel, and special demolition of military fortifications. During a three-month period in 2007 alone, they conducted missions with two U.S. Army military police brigades in MND-N, locating and disposing of over 20,300 rounds of unexploded ordnance. During their five years of service in Iraq, Moldovan EOD personnel destroyed over 400,000 rounds of ammunition. With an increasing number of Iraqi Security Forces trained in EOD operations, Moldova handed over those duties to the Iraqis and withdrew its forces from Iraq in December 2008.

MONGOLIA

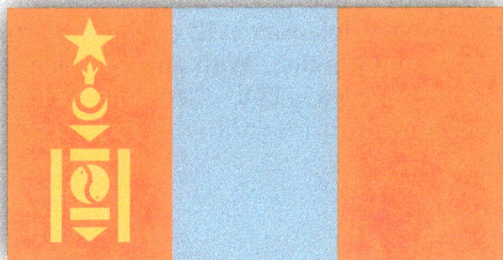

Ground troops deployed (cumulative): 1,128
Peak deployment: 180
Deployment dates: September 2003–September 2008
Unit designation: Desert Lions
Order of battle: MND-CS
Lead: Poland
Primary deployment location(s): Camp Charlie, Al Hillah; Camp Echo
Casualties (dead): 0
Casualties (wounded): 0

Force overview: The first Mongolian contingent of 180 troops arrived in Iraq in September 2003. The force was assigned to the Polish multinational division and initially deployed to Camp Charlie at Al Hillah in Babil Province. The Mongolian forces moved to Camp Echo in Ad Diwaniyah in 2006, where they remained until they withdrew from Iraq in September 2008. The contingent was reduced to one hundred personnel in 2007 and remained at that level until September 2008. The force included an infantry company, military engineers, and medical teams.

Operations: Mongolian troops were deployed in Iraq for over five years, generally in 100-man infantry companies. Their initial mission was providing security for Camp Charlie, located in the city of Al Hillah, south of Baghdad. In February 2004, Mongolian soldiers on guard duty at Camp Echo shot and killed an insurgent attempting to drive a vehicle loaded with explosives onto the base. Because of the soldiers' rapid response, the driver did not penetrate the checkpoint and the explosives detonated early, limiting the number of casualties in the attack.

THE NETHERLANDS

Ground troops deployed (cumulative): 7,564
Peak deployment: 1,345
Deployment dates: 1 August 2003–7 March 2005
Unit designation: Battlegroup Stabilization Force Iraq (SFIR)
Order of battle: MND-SE
Lead: United Kingdom
Primary deployment location(s): As Samawah, Al Muthanna Province
Casualties (dead): 2
Casualties (wounded): Unknown

Force overview: The Netherlands deployed military forces to Iraq in August 2003 under the auspices of the United Nations–sponsored Stabilization Force Iraq. Designated as Battlegroup Stabilization Force Iraq, the Dutch soldiers served in the British-led MND-SE. The joint SFIR consisted of 1,345 personnel at its peak, including contingents from the Royal Netherlands Army forces (a commando squad, logistics team, and field hospital), Marines (part of the Royal Netherlands Navy), Air Force (Chinook, Cougar, and Dutch Apache helicopters), and military police (the independent Royal Marechaussee, or Royal Constabulary). The Dutch deployed to three main locations within the province of Al Muthanna: As Samawah, Ar Rumaythah, and Al Khidr. Its rotary-wing assets were located at Tallil Air Base. The Netherlands also positioned military personnel at the coalition headquarters in Al Basrah and Baghdad.

Operations: As Samawah, Al Muthanna Province, had been under American control since the 82d Airborne Division captured the city in March 2003. It was later secured by the U.S. 1st Marine Division. The Dutch SFIR took possession of the region from the U.S. marines in August 2003. Its mission was to rebuild the local Iraqi infrastructure and to supply security for the region, including the archeological site at Uruk (also known as Erech). Supporting tasks included training and equipping Iraqi Security Forces, mentoring Iraqi government officials, and rebuilding the economy, public utilities, and infrastructure. Initially the Dutch SFIR focused on restoring public utilities, including water, sewage, electricity, gas, and communications.

Beginning in January 2004, the Dutch force also became responsible for protecting Japanese forces in As Samawah, since the Japanese rules of engagement prevented them from conducting their own security. Nevertheless, the security situation in As Samawah appeared stable, with few signs of strife. In order to prepare for the transfer of sovereignty to Iraq, initially scheduled for June 2004, the Dutch SFIR focused on building up sufficient security forces and security infrastructure during the first half of the year. It helped establish the Provincial Joint Coordination Center in As Samawah, where the Dutch worked continually with local Iraqi officials and security forces, training some 2,800 Iraqi soldiers and police. Continuous daylight and nighttime patrols enabled the Dutch soldiers to maintain public order and interact with the population.

Initial "kinetic" operations focused on stopping looters who targeted fuel dumps along main supply routes Jackson and Tampa. In April and May 2004, however, the force faced an insurrection among followers of Moqtada al-Sadr that resulted in increased mortar attacks and ambushes. As a result, the battlegroup was reinforced with three counterbattery radars to locate insurgent mortar positions and six Dutch Apache helicopters to provide quick responses. The situation seemed to stabilize in June and

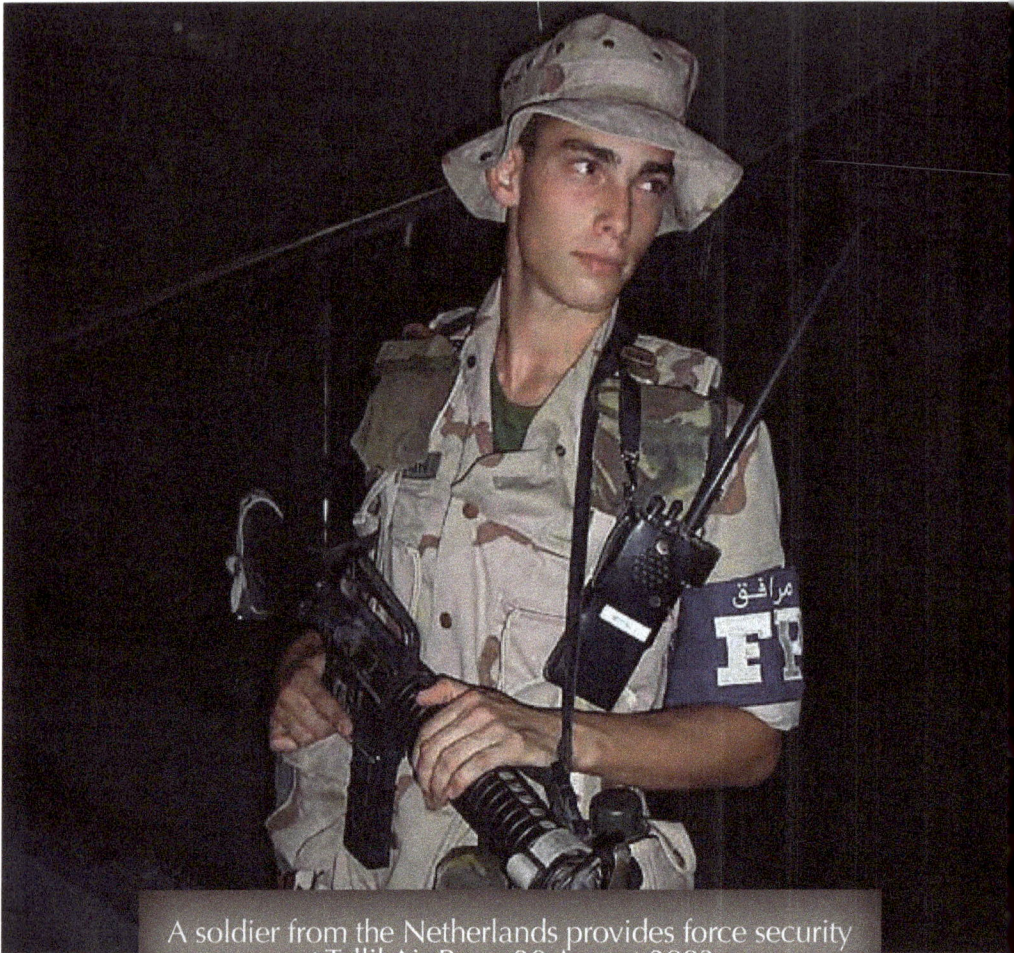

A soldier from the Netherlands provides force security
at Tallil Air Base, 28 August 2003.

July, after sovereignty was returned to the Iraqis but flared up in
August when insurgents attacked a Dutch military police patrol in
the city of Ar Rumaythah, killing one warrant officer. The rapid-
response force dispatched to relieve the patrol also came under
heavy insurgent fire but managed to reach the besieged units and
establish a defensive perimeter. The attackers fled when the Apache
gunships arrived. The SFIR command responded to the attack with

increased patrols in the major provincial cities and redoubling intelligence gathering. The battlegroup was also reinforced with two additional infantry platoons. With the additional manpower, the Dutch increased the number of patrols as well as the number of civil-military projects, and by November 2004 the security situation in the province returned to normal.

When the Netherlands' forces withdrew from Al Muthanna in March 2005, the United Kingdom and Australia provided security for a short time in the province before turning full control over to the Iraqi government. Al Muthanna was the first province to be fully independent and responsible for its own security, with much of the credit going to the Netherlands Stabilization Force during its nineteen-month deployment.

Other military contributions: The Netherlands also provided six instructors to train the Iraqi coast guard (Iraqi Coastal Defense Force) in Umm Qasr as part of a larger group of eighty-five instructors, primarily from the United Kingdom.

The Netherlands also provided personnel to the NATO-sponsored operation to train the Iraqi Security Forces' noncommissioned officers and officers.

NEW ZEALAND

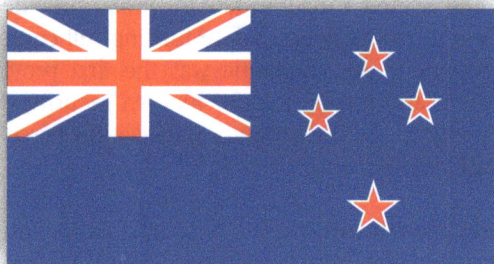

Ground troops deployed (cumulative): 250
Peak deployment: 161
Deployment dates: September 2003–September 2004
Unit designation: Not applicable
Order of battle: MND-SE
Lead: United Kingdom
Primary deployment location(s): Al Basrah
Casualties (dead): 0
Casualties (wounded): 0

Force overview: New Zealand responded quickly to the humanitarian needs of the Iraqi people following the conclusion of major combat operations. After the UN Security Council passed Resolution 1483 on 22 May 2003, New Zealand sent sixty-one military engineers to work alongside British forces in southern Iraq, near Al Basrah. The engineers completed their mission in Iraq and left the country in September 2004. The initial deployment also included two New Zealand Defense Force personnel as part of the UN Mine Action Service operation.

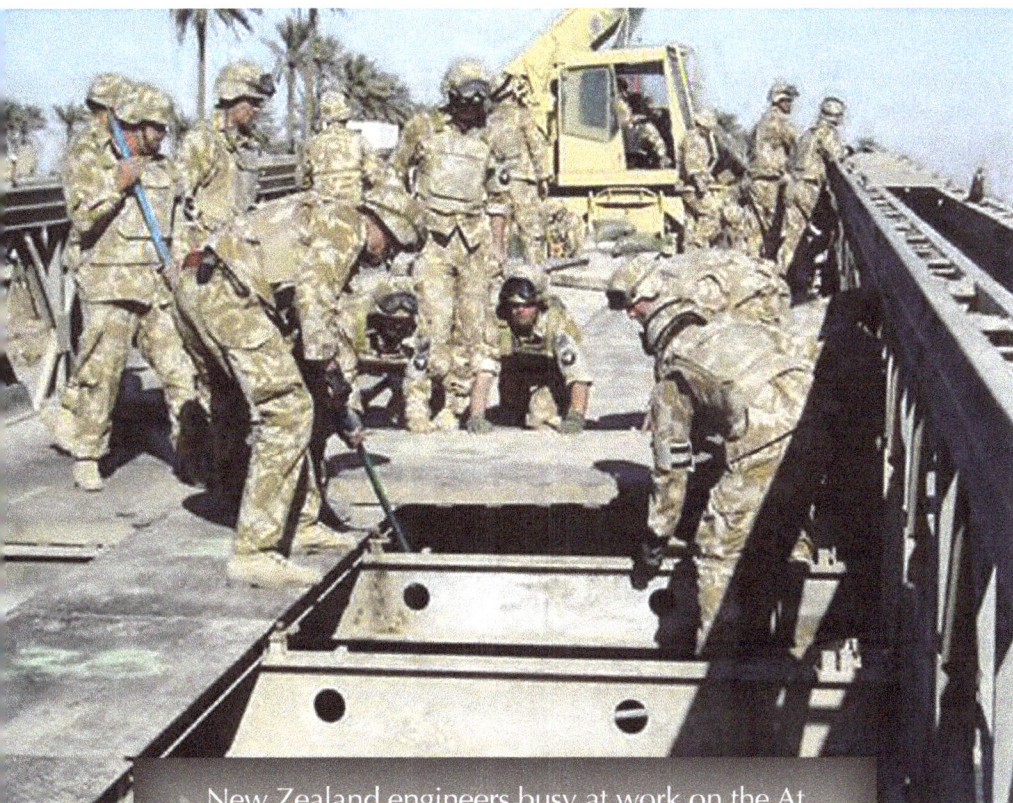

New Zealand engineers busy at work on the At Tannumah Bridge, Al Basrah, October 2003

Operations: The New Zealand engineers were responsible for repairing and maintaining water systems, power facilities, bridges, and other infrastructure around Al Basrah. Specifically, the New Zealand engineers built two bridges, the Cullingworth and At Tannumah, connecting Al Basrah to its northeast bank across the Shatt al-Arab, the major river that runs through the city. These structures provided access to the main trade routes between the rural villages and Al Basrah's markets. The Linton's 2d Engineer Regiment also worked with the U.K.'s 77th Armoured Engineer Squadron to construct a reverse osmosis water plant in At Tannumah, supplying over 200,000 people around Al Basrah with fresh drinking water. Finally, the engineers also installed fresh-water tanks at schools throughout Al Basrah, so that students could have regular access to fresh water.

Other military contributions: New Zealand has one military liaison officer attached to the UN Assistance Mission for Iraq.

Nicaragua

Ground troops deployed (cumulative): 115
Peak deployment: 115
Deployment dates: September 2003–February 2004
Unit designation: Part of Multi-National Brigade Plus Ultra
Order of battle: MND-CS
Lead: Poland (under direct Spanish control)
Primary deployment location(s): Base España, Ad Diwaniyah
Casualties (dead): 0
Casualties (wounded): 0

Force overview: In September 2003, Nicaragua deployed 115 soldiers to serve in the Plus Ultra Brigade. The Nicaraguan troops were stationed in Ad Diwaniyah at Base España, along with the bulk of the large Spanish military contingent. Most of the troops in Nicaragua's detachment were trained sappers with experience in disarming and destroying unexploded ordnance.

Operations: The primary mission of the Nicaraguan military engineers was to disarm unexploded artillery. During their single deployment, the sappers destroyed over twenty-two thousand tons of explosives and provided expertise to other coalition troops regarding Soviet-era weapons and munitions.

Due to a lack of funding, the contingent withdrew in February 2004, well before Spain and several of the other nations in the Plus Ultra Brigade.

NORWAY

Ground troops deployed (cumulative): 300
Peak deployment: 150
Deployment dates: July 2003–June 2004
Unit designation: Not applicable
Order of battle: MND-SE
Lead: United Kingdom
Primary deployment location(s): Al Basrah
Casualties (dead): 0
Casualties (wounded): 0

Force overview: Norway deployed a company of 150 military engineers and bomb disposal experts in July 2003. The contingent deployed to Al Basrah, supervised by MND-SE, and the troops were recalled in June 2004.

Operations: The primary responsibility of Norwegian forces in Iraq was to dispose of unexploded ordnance. In addition, they provided engineer services.

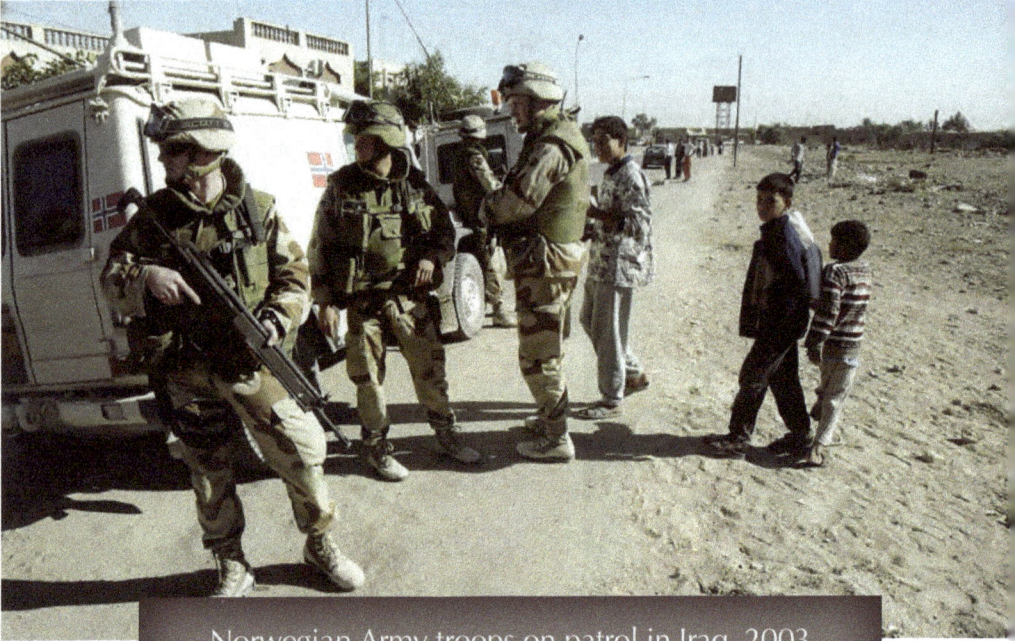

Norwegian Army troops on patrol in Iraq, 2003

Other military contributions: Norway sent ten liaison officers to serve at MND-SE and MND-CS from June 2005 to August 2006. Ten Norwegian military instructors were sent to help train Iraqi military in December 2004 under the NATO Training Mission–Iraq.

Philippines

Ground troops deployed (cumulative): 100
Peak deployment: 51
Deployment dates: July 2003–July 2004
Unit designation: Not applicable
Order of battle: MND-CS
Lead: Poland
Primary deployment location(s): MND-CS
Casualties (dead): 0
Casualties (wounded): 0

Force overview: The Philippines deployed fifty-one soldiers to Iraq in July 2003. The contingent included engineers, as well as medical and security personnel. The soldiers were assigned to the Polish multinational division, which became the core of MND-CS.

Operations: The Filipino contingent's mission was to provide humanitarian relief. Its forces also trained Iraqi police and military personnel, patrolled with U.S. and Polish military police, and conducted extensive medical outreach programs. In July 2004, Filipino President Gloria Macapagal-Arroyo recalled the soldiers after a Filipino contractor had been kidnapped by insurgents and threatened with beheading if the Philippines did not withdraw.

POLAND

Ground troops deployed (cumulative): 13,900
Peak deployment: 2,400
Deployment dates: March 2003–October 2008
Unit designation: Not applicable
Order of battle: MND-CS
Lead: Poland
Primary deployment location(s): Karbala; Ad Diwaniyah
Casualties (dead): 23
Casualties (wounded): Unknown

Force overview: Poland was one of only five nations to have ground forces involved in major combat operations, with its troops participating in the initial invasion. The first contingent consisted of 125 soldiers from the elite GROM commando unit and 24 "Marine Division," along with 74 chemical and biological warfare troops from 4 Brodnicki Pułk Chemiczny. In August 2003, a new contingent of 2,400 mechanized infantry soldiers replaced the original units. During their remaining tenure in Iraq, Polish rotations alternated between mechanized infantry

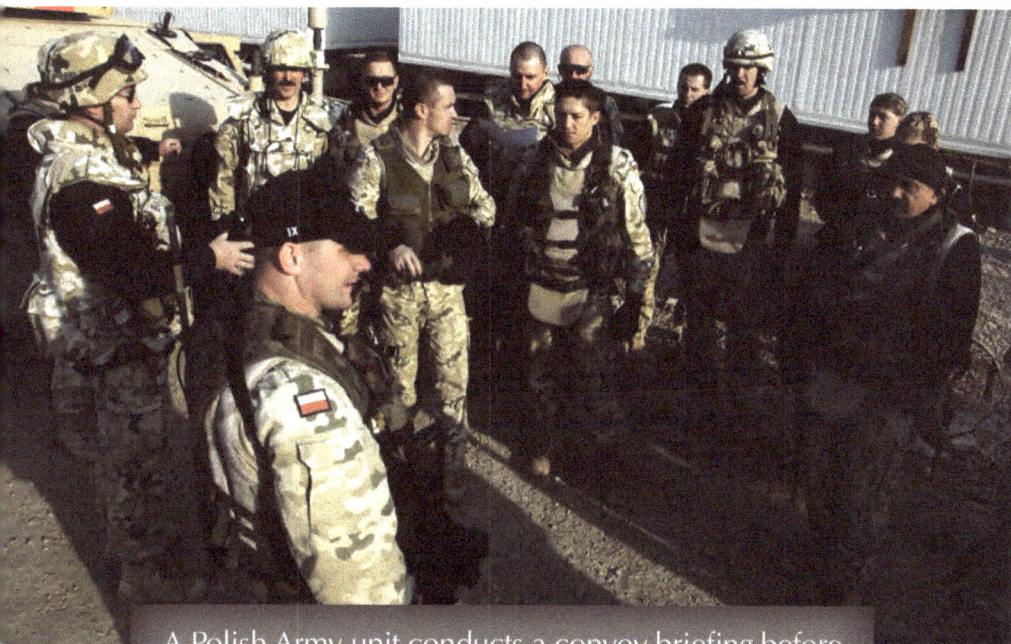

A Polish Army unit conducts a convoy briefing before heading out on a mission from Camp Echo, 30 December 2007.

and cavalry units. In 2005 Poland reduced its troop strength to 1,500 and in 2006 to 900 soldiers. The contingent remained at that level until Poland withdrew in October 2008.

From August 2003 to the end of its deployment, Poland commanded multinational forces in southern Iraq that were subsequently dubbed Multi-National Division–Center-South. Initially, the area of Polish responsibility included Babil, Karbala, Wasit, An Najaf, and Al Qadisiyah Provinces. After the creation of MND-CS in 2004, the area was reduced to the provinces of Babil, Wasit, and Al Qadisiyah. The biggest challenge was integrating troops from twenty-one nations into a cohesive force. MND-CS included troops from Armenia, Bulgaria, Denmark, El Salvador, Kazakhstan, Latvia, Lithuania, Mongolia, Poland, Romania, Slovakia, and the Ukraine, all with different specializations, equipment, and doctrines, and all with different instructions from their governments that limited what the forces could do.

Beginning with its deployment in August 2003, the Polish force was composed of a Polish Brigade Combat Team, which included an infantry maneuver group, aviation group, military police, and other support units.

Operations: During initial combat operations, Polish commandos took part in security operations to prevent the destruction of Iraqi oil platforms. Polish special forces also assisted in securing the port of Umm Qasr. With the establishment of the Polish multinational division in August 2003, the Polish contingent and the various coalition forces assigned to MND-CS provided security to the Iraqi inhabitants in their zone, restored infrastructure and facilities, delivered humanitarian aid, and trained units of the Iraqi Army and police. In 2007 alone, the Polish contingent was involved in over 227 reconstruction projects.

Polish leadership and operations allowed the provinces under MND-CS to be rebuilt quickly and returned to Iraqi control. When Polish forces withdrew in October 2008, the significance of their contributions showed in many ways, including the high quality of the Iraqi forces they helped train.

PORTUGAL

Ground troops deployed (cumulative): 256
Peak deployment: 128
Deployment dates: October 2003–February 2005
Unit designation: Not applicable
Order of battle: MND-SE
Lead: United Kingdom (served under Italian Carabinieri)
Primary deployment location(s): An Nasiriyah
Casualties (killed): 0
Casualties (wounded): 0

Force overview: In October 2003, Portugal deployed 128 military police to An Nasiriyah, where they were stationed along with Italian Carabinieri personnel. Despite domestic pressure to withdraw the force at the end of its initial twelve-month deployment, Portugal maintained its military presence in Iraq until the January 2005 elections, after which it withdrew.

Other military contributions: From early 2006 to 2009, Portugal maintained a team of six to eight trainers as part of the NATO Training Mission–Iraq.

101

REPUBLIC OF KOREA

Ground troops deployed (cumulative): 20,000
Peak deployment: 3,600
Deployment dates: April 2003–December 2008
Unit designation: Zaytun Division
Order of battle: MND-NE
Lead: Republic of Korea
Primary deployment location(s): Arbil
Casualties (dead): 1
Casualties (wounded): Unknown

Force overview: The Republic of Korea initially dispatched a contingent of 600 military medics and engineers (known as the Seohee and Jema units, respectively) to southern Iraq in April 2003. In September 2003, responding to a U.S. request, Korea agreed to increase its contingent, sending the newly formed Zaytun Division (*zaytun* is Arabic for olive). This division, composed initially of the 11th and 12th Infantry Brigades as well as division support forces, eventually deployed some 2,200 troops, including engineers and security forces, to Arbil in the Kurdish Autonomous Region of northern Iraq in September 2004. The addition of the Seohee

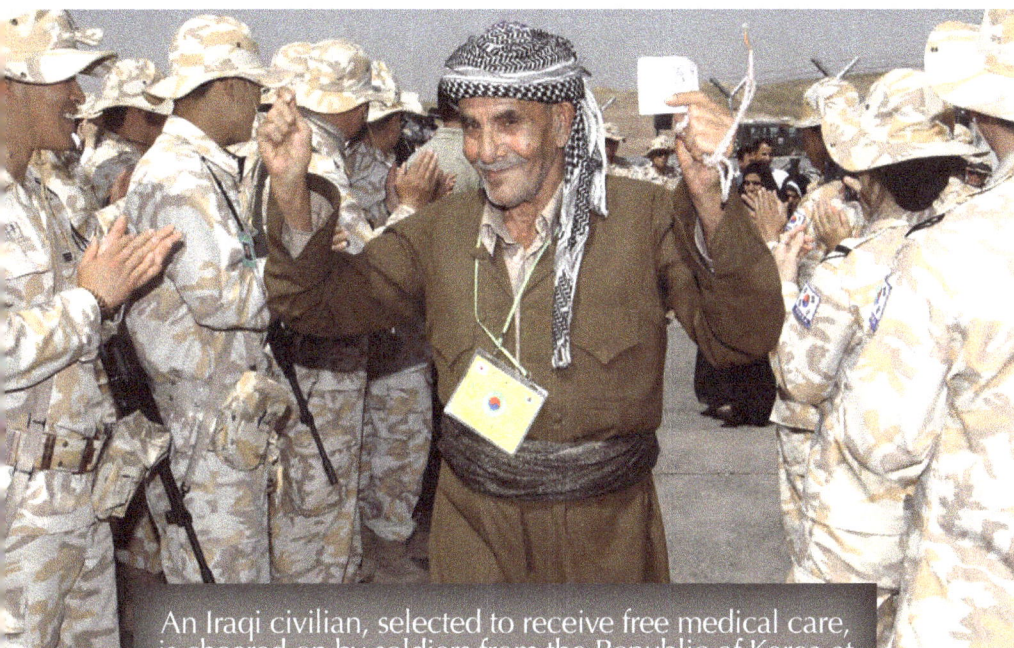

An Iraqi civilian, selected to receive free medical care, is cheered on by soldiers from the Republic of Korea at Camp Zaytun, Arbil, 24 May 2005.

and Jema medical units brought the division's strength to 2,800 soldiers. Another 800 soldiers were dispatched to reinforce the existing troops in Arbil in November 2004, further increasing the size of the Korean contingent to 3,600.

In March 2006, Korea withdrew 1,300 troops, followed by another 1,200 troops in March 2007, dropping the strength of the Korean contribution to 1,100. By December 2007, the Korean force was reduced to six hundred personnel, where it remained until the withdrawal of its forces at the end of 2008.

Operations: The main mission of the Korean troops was humanitarian. Their tasks included providing medical services and building and repairing roads, schools, power lines, and other public infrastructure. Under MND-NE, the Zaytun Division worked closely with the Kurdish Regional Government and provided security and escort for UNAMI personnel. The contingent also aided the Iraqi Security Forces by supplying facilities, equipment, and training; operating the Zaytun Vocational Training Center to conduct reconstruction support projects; and assisting local education and medical care facilities. It also supported the Zaytun hospital by providing medical supplies and equipment, internship training programs for local doctors and nurses, and mobile clinic services to remote towns. The Koreans treated over forty thousand Iraqis in their five-year engagement with Iraq.

ROMANIA

Ground troops deployed (cumulative): 6,600
Peak deployment: 730
Deployment dates: July 2003–23 July 2009
Unit designation: Black Wolves, Red Scorpions, Carpathian Hawks
Order of battle: MND-B; MND-CS; MND-SE
Lead: United States; Poland; United Kingdom
Primary deployment location(s): An Nasiriyah; Tallil
Casualties (dead): 3
Casualties (wounded): Unknown

Force overview: In July 2003, Romania deployed one infantry battalion with 550 soldiers to serve as part of the Italian Garibaldi Brigade in An Nasiriyah. In November 2004, the force grew to 730 personnel. While at that level, the Romanian force consisted of roughly 400 infantrymen, 100 military police, 150 EOD specialists, 30 medics, and 50 intelligence officers. At the peak of their deployment, Romanian soldiers were assigned to MND-B, MND-CS, and MND-SE. In July 2006, Romanian forces were reduced to about six hundred personnel. Romania formally terminated its Iraq

105

The color guard for Romania's 26th Infantry Battalion marches during the end-of-mission ceremony at Camp Dracula, 4 June 2009.

mission on 4 June 2009, and its last troops left the country on 23 July 2009.

Operations: Because Romanian forces were deployed to three multinational divisions simultaneously, their missions were diverse. They provided intelligence support to MND-C and MND-CS by conducting human intelligence gathering, interrogations, and reconnaissance and surveillance, as well as operating unmanned aerial vehicles in order to support near-real-time intelligence. In MND-SE, Romanian forces provided security to the base at Tallil and the major supply routes, as well as serving as a quick-reaction force at Tallil. In addition, the Romanians conducted mechanized combat patrols and bridge security. Finally, Romanian Army medics ran a hospital providing emergency medical treatment for detainees and military personnel at Camps Cropper and Bucca.

Like other coalition contingents, Romanian troops undertook many civil works projects, including refurbishing schools and supporting Iraqi hospitals, in addition to rebuilding water-sanitation facilities, pumping stations, and communication lines.

SLOVAKIA

Ground troops deployed (cumulative): 425
Peak deployment: 85
Deployment dates: 8 June 2003–February 2007
Unit designation: Not applicable
Order of battle: MND-CS
Lead: Poland
Primary deployment location(s): Ad Diwaniyah
Casualties (dead): 4
Casualties (wounded): Unknown

Force overview: On 8 June 2003, eighty-five Slovakian military engineers were sent to aid multinational forces in Iraq. All were stationed at Camp Echo, near Ad Diwaniyah, with MND-CS.

Operations: Slovakia's primary military mission involved anti-mining operations. Its engineers performed survey reconnaissance, removal and demolition of unexploded ordnance, manual and mechanical de-mining, ground excavation, and construction of fortifications. Slovakia redeployed the engineering unit home in February 2007 after seven rotations.

In total, during their time in Iraq, Slovakian engineers cleared 2.5 million square meters of land, cleared 850,000 square meters of transportation routes, manually de-mined 1 million square meters of land, destroyed 1 million pieces of ordnance, and detonated 700 tons of other explosives.

Eight staff officers remained with Multi-National Force–Iraq and Multi-National Corps–Iraq after February 2007, but Slovakia eventually withdrew its last liaison officers from Camp Victory, Baghdad, in December of that year.

SPAIN

Ground troops deployed (cumulative): 4,100
Peak deployment: 1,300
Deployment dates: 21 March 2003–23 April 2004
Unit designation: Multi-National Brigade Plus Ultra
Order of battle: MND-CS
Lead: Poland
Primary deployment location(s): Ad Diwaniyah; An Najaf
Casualties (dead): 11
Casualties (wounded): Unknown

Force overview: Spain deployed approximately nine hundred troops to Iraq shortly after the start of the conflict. The original contingent focused on humanitarian missions, and the soldiers were generally medical personnel, engineers, and logistical units. After the troops' first rotation, however, the Spanish government decided to send a more robust combat force of 1,300 troops, including contingents from the Spanish legionnaires (2d Spanish Legion Brigade) and special operations forces (from the Mando de Operaciones Especiales).

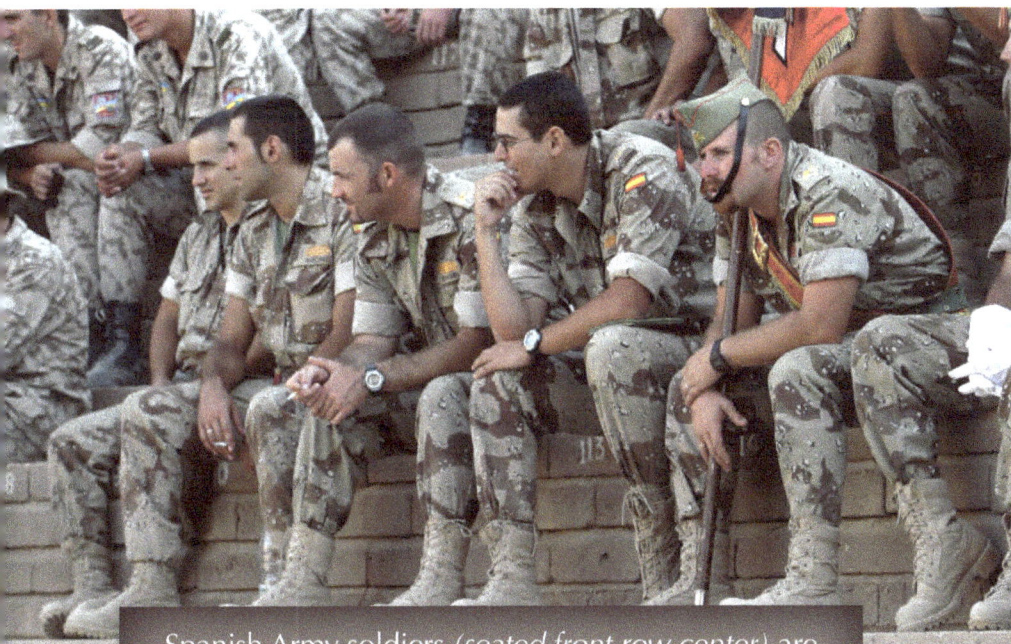

Spanish Army soldiers *(seated front row center)* are among the multinational coalition forces watching a ceremony at Camp Babylon, 3 September 2003.

Beginning in July 2003, the Spanish military leader also commanded what became known as the Plus Ultra Brigade, or Brigada Hispanoamericana. The brigade consisted of Spain's contingent plus soldiers from the Dominican Republic, El Salvador, Honduras, and Nicaragua. The Central American countries contributed 1,200 soldiers, so in total the Plus Ultra Brigade had some 2,500 troops at its peak.

Multi-National Brigade Plus Ultra had five operational bases, two of which were located in Ad Diwaniyah Province while the other three were in An Najaf Province. In Ad Diwaniyah, Base España was the brigade's command post and housed a majority of the Spanish and Nicaraguan contingents. The other, Base Santo Domingo, was manned by forces from the Dominican Republic. In An Najaf, Base El Salvador was home to the Salvadoran troops, while Base Tegucigalpa accommodated the Honduran soldiers.

111

The Al Andalus base was the site of MNB PU's alternate command post and the An Najaf Coalition Provisional Authority.

Operations: The Spanish contingent focused on providing humanitarian relief, medical aid, infrastructure repair, community outreach, and security in south-central Iraq.

Plus Ultra's area of responsibility was the provinces of Al Qadisiyah and An Najaf, two largely Shi'ite regions, each with about 1 million inhabitants.

During its deployment, the brigade's five camps were hit by mortar and other small arms attacks. In an ambush on 11 November 2003, five Spanish soldiers were killed in action.

The Plus Ultra Brigade dissolved in April 2004. At that time, the newly elected Spanish prime minister had ordered Spain's troops home as rapidly as possible in response to the lack of public support and in the aftermath of the Madrid train bombing in March 2004. The other members of Plus Ultra withdrew at the same time, except for El Salvador, which sustained its combat presence in Iraq until 2009.

THAILAND

Ground troops deployed (cumulative): 866
Peak deployment: 433
Deployment dates: 30 September 2003–30 September 2004
Unit designation: Thai Humanitarian Assistance Task Force 976–
Iraq
Order of battle: MND-CS
Lead: Poland
Primary deployment location(s): Camp Lima, Karbala
Casualties (dead): 2
Casualties (wounded): 5

Force overview: Thailand's Task Force 976–Iraq consisted of an engineer battalion, six medical service teams, a force security platoon, and a support platoon. The task force served under Polish command in MND-CS.

Operations: The Thai force deployed to Camp Lima in Karbala in September 2003 to provide engineering support, civil-military operations, and humanitarian assistance. It rebuilt local hospitals and clinics, renovated and reopened schools, and repaired other

113

infrastructure facilities. The Thai engineers also assisted in constructing and repairing MND-CS installations around Karbala, while the medical service teams administered medical care to locals and provided physicians to support the Polish medical company.

On 27 December 2003, suicide bombers struck Camp Lima, some 100 kilometers southwest of Baghdad, along with another coalition installation and an Iraqi police station (all in Karbala). The Camp Lima attacker rammed his vehicle into the post's wall, killing two guards from the Thai security platoon and wounding five other Thai soldiers. In all, six coalition troops were killed (the other four being Bulgarians) and thirty-seven coalition troops were wounded. At least six civilians were killed and many more wounded. The attack was atypical, as a majority of the violence at the time was centered on the Sunni Triangle area around Baghdad. Despite domestic pressure to remove the troops after the attack, the Thai government decided to keep the task force deployed until its original mandate expired on 30 September 2004.

TONGA

Ground troops deployed (cumulative): 200
Peak deployment: 55
Deployment dates: 13 June–December 2004; 18 August–November 2008
Unit designation: Not applicable
Order of battle: MND-B
Lead: United States
Primary deployment location(s): Camp Blue Diamond, Ar Ramadi; Al Faw Palace, Baghdad
Casualties (dead): 0
Casualties (wounded): 0

Force overview: During their first Iraq deployment in 2004, forty-five Tongan Royal Marines augmented the U.S. I Marine Expeditionary Force in Al Anbar Province, performing security duties at Camp Blue Diamond. With their second deployment in 2007, fifty-five Tongan Royal Marines provided security for the multinational forces headquarters at Al Faw Palace, Baghdad.

Operations: At its peak deployment in Iraq, the Tongan contribution amounted to over 10 percent of that country's total military force of 450 soldiers.

UKRAINE

Ground troops deployed (cumulative): 7,000
Peak deployment: 1,630
Deployment dates: March 2003–22 December 2005
Unit designation: Not applicable
Order of battle: MND-CS
Lead: Poland
Primary deployment location(s): Al Kut, Wasit Province
Casualties (dead): 18
Casualties (wounded): 32

Force overview: The initial Ukrainian deployment consisted of 448 soldiers of a chemical, nuclear, and biological warfare battalion to support the invasion of Iraq in March 2003. They were replaced on 28 August 2003 by 1,621 troops from the 5th Separate Mechanized Brigade, followed by rotations of the 6th and 7th Separate Mechanized Brigades. All served with MND-CS around Wasit Province. The Ukrainian force began to draw down on 7 May 2005 when the 7th Brigade was replaced by the 81st Tactical Group, dropping the force to nine hundred infantrymen. During this final deployment, the Ukrainian force slowly reduced its troop

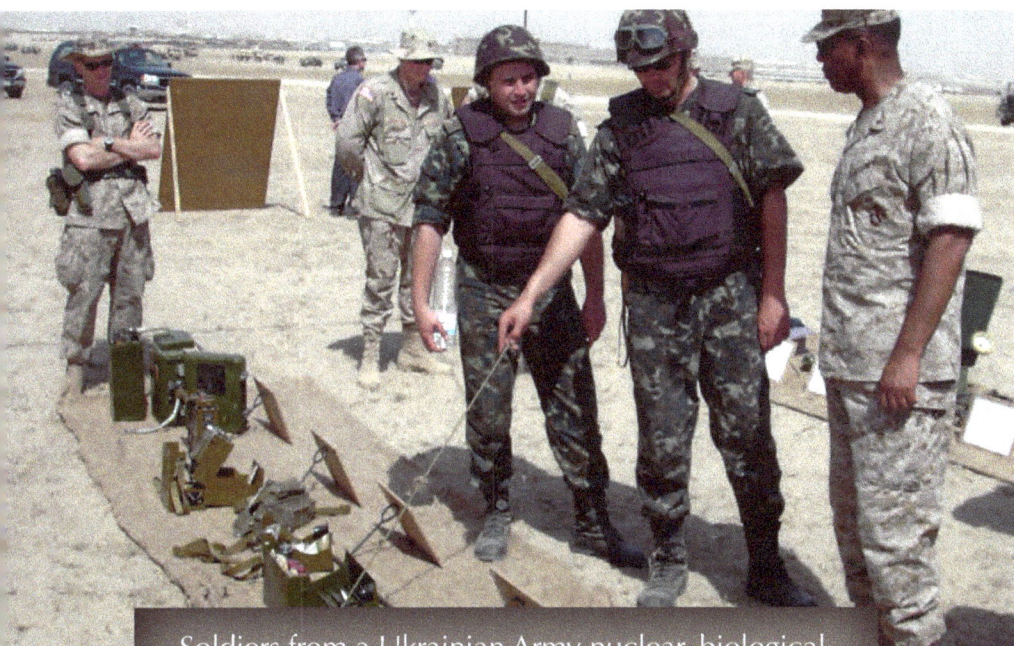

Soldiers from a Ukrainian Army nuclear, biological, and chemical unit discuss their techniques and equipment with coalition partners at Camp Arifjan, Kuwait, 3 August 2003.

commitment, with its final forty-four troops leaving Iraq on 22 December 2005.

Operations: The first Ukrainian forces were chemical and biological weapons specialists in case Iraq employed those weapons during the initial invasion. The Ukrainian troops that deployed to Iraq provided security to convoys and checkpoints, as well as local government and religious officials. They also protected bridges and historical sites.

During their security missions, Ukrainian forces apprehended some fourteen thousand suspects, seized nine hundred weapons and thirteen thousand rounds of ammunition, and took part in disposing of some two million rounds of ammunition. Additionally, they trained three Iraqi infantry brigades and six thousand Iraqi

117

military and police personnel. The Ukrainian contingent also participated in many larger combined-combat operations, such as CASCADE (to detect and seize the leadership of the Al-Sadr party); CHAMBERLAIN (to block illegal immigration and trafficking of weapons and drugs across the border); and ELECTION 2005 (to secure the area of responsibility during Iraqi elections). Ukrainian soldiers also prepared and conducted joint operations FAIRWATER, CORRIDOR, and BORDER with Iraqi armed forces, police, and border patrol along the Iraq-Iran border.

Other military contributions: Ukraine continues to provide personnel to the NATO Training Mission–Iraq and United Nations Assistance Mission for Iraq.

United Kingdom

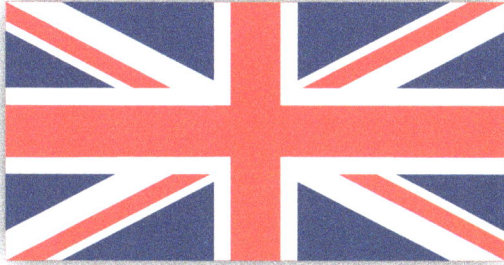

Ground troops deployed (cumulative): 102,000
Peak deployment: 46,000
Deployment dates: 20 March 2003–28 July 2009
Unit designation: Not applicable
Order of battle: MND-SE
Lead: United Kingdom
Primary deployment location(s): Al Basrah
Casualties (dead): 179
Casualties (wounded): 1,747

Force overview: Operation TELIC (literally "purposeful action") was the code name of all British operations in Iraq. About 46,000 troops from the British military branches were committed to TELIC I, which encompassed the March invasion through the end of major combat operations in June 2003. During this phase, some 26,000 British Army soldiers, 4,000 Royal Marines, 5,000 Royal Navy and Royal Fleet Auxiliary sailors, and 8,100 Royal Air Force airmen were actively involved in the operation. After the end of TELIC I, the United Kingdom made twelve rotations, or *roulements* of troops (a term used by the British Army to signify deployments of major combat units).

119

The British 1st Armoured Division headed TELIC I, a force of three army brigades: the 16th Air Assault Brigade, the 7th Armoured Brigade, and the 102d Logistics Brigade. The Royal Marines 3d Commando Brigade was also under the operational command of the division.

Additional rotations began shortly after the declared end of major combat operations in 2003. The 3d Commando Brigade and the 16th Air Assault Brigade largely withdrew in May, and the 101st Logistics Brigade relieved the 102d Logistics Brigade in late May. The 19th Mechanised Brigade relieved the 7th Armoured Brigade in June 2003, and the 3d Division replaced the 1st Armoured Division on 11 July 2003, signaling the start of TELIC II. The 3d Division controlled numerous other coalition forces in southeast Iraq, including contingents from Italy, the Netherlands, Denmark, the Czech Republic, Lithuania, Norway, and New Zealand. (See *Table 2* for additional rotations.)

The United Kingdom officially terminated combat operations on 30 April 2009, with the last of its troops leaving Iraq on 28 July 2009.

Operations: During their nearly six-year involvement in Operation TELIC, U.K. forces performed diverse roles and missions in support of coalition efforts. The initial mission provided full-spectrum forces

Table 2—U.K. Unit Rotations, 2003–2007

Date	Units in Theater
November 2003	19th Mechanised Bde
April 2004	20th Armoured Bde
October 2004	1st Mechanised Bde
May 2005	4th Armoured Bde
October 2005	7th Armoured Bde
April 2006	20th Armoured Bde
June 2007	1st Mechanised Bde
December 2007	4th Mechanised Bde
June 2008	7th Armoured Bde

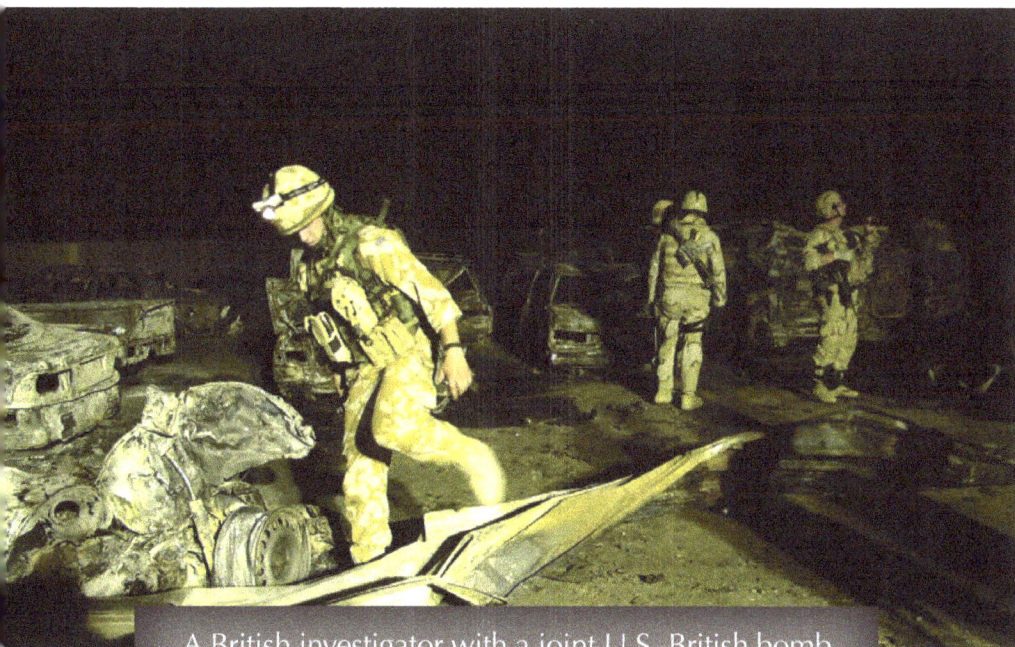

A British investigator with a joint U.S.-British bomb examination team studies the wreckage from a car bombing near Camp Rashid, 1 August 2004.

for the liberation of Iraq. With the end of major combat, the British shifted their focus to help the Iraqis secure and rebuild their country after years of neglect and conflict. The United Kingdom supported reconstruction in southeast Iraq, assisting the Iraqis build their economy and have an economic stake in the future. U.K. forces also helped train the new Iraqi Army and other security forces.

Aside from the United States and the current Iraqi Security Forces, the United Kingdom consistently contributed more resources and troops to the Iraq mission than any other coalition partner. U.K. forces took part in many operations that are far beyond the scope of this study. Of special interest, however, is the United Kingdom's role during TELIC I and its contribution to the early success of the major combat phase of the conflict.

The land offensive began on 20 March 2003, less than twenty-four hours after the air campaign started. The 1st Armoured

Division's initial objective was to seize the Al Faw peninsula and secure the vitally important port of Umm Qasr. Simultaneously, forty Commando (Royal Marines) and other coalition forces launched an amphibious helicopter assault to seize key Iraqi oil infrastructure on the Al Faw peninsula. The assault was supported by the U.K. Joint Helicopter Command, a variety of landing craft, and three Royal Navy frigates providing fire support.

The securing of the Al Faw peninsula and the Rumaylah oil fields in southeast Iraq by U.K. and U.S. forces allowed coalition personnel to race north. The United Kingdom's attention then turned to securing Al Basrah, Iraq's second-largest city, to prevent Iraqi forces from staging attacks on coalition logistical lines of communications. Despite encountering significant Iraqi resistance, U.K. forces seized Al Basrah's airport in four days and began expanding their area of control in the surrounding region.

After several days of probes, U.K. forces took the town of Az Zubayr, southwest of Al Basrah, and on 6 April 2003 entered Al Basrah in strength. After encountering pockets of resistance, the forces stormed the Ba'ath Party headquarters. Although there was some initial looting in the town, the situation quickly returned to normal. The U.K. commanders established contact with local leaders and assisted in restoring a functioning police force. The first joint U.K.-Iraqi police patrols took place just one week after Al Basrah had been liberated. The British success in southeast Iraq enabled U.S. troops to push swiftly toward Baghdad and take control of most of that city between 8 and 9 April 2003.

After conducting thousands of operations aimed at securing the region in the years since major combat operations ended, in 2007 British forces changed focus to assist and advise the Iraqi Security Forces. The British presence in southeast Iraq, however, was not without its difficulties. In early 2007, the British were preparing to hand Al Basrah over to Iraqi Security Forces. By February 2007, the British maintained only two bases in the city. The largest contingent of some five thousand soldiers was located at the Al Basrah airport on the outskirts of town. The only contingent left inside the city was a small force of seven hundred soldiers at the Al Basrah palace. Beginning on 27 February, several Iraqi enemy militia forces, including the Mahdi Army (followers of Shi'ite cleric Moqtada al-Sadr), began operations against British forces in the area. The well-organized insurgent tactics included mortar and rocket attacks against the Al Basrah airport, as well as ambushes and IED attacks against British patrols and convoys that attempted to resupply the Al Basrah palace garrison. In short, the insurgents prevented the British from launching anything more than limited raids and critical resupply missions and essentially grounded most U.K. rotary-wing aircraft for fear of losing them to heavy ground fire.

As the situation worsened, British forces abandoned their Al Basrah palace garrison on the night of 3 September and withdrew to their larger airport base. By doing so, they ceded control of the city to insurgents. Insurgent attacks then turned from targeting the British and focused on Iraqi police and other militias, leading to complete lawlessness in the city. Finally, in March 2008, Iraqi Army forces began major operations in Al Basrah to regain control of the area.

Although the Al Basrah situation between 2007 and 2008 represented a major setback during the British engagement in Iraq, the United Kingdom's other successes and contributions to the conflict were significant. The U.K. presence generally allowed U.S. forces to employ their combat power elsewhere in Iraq, and the United Kingdom was quickly able to secure and prepare several key provinces for their return to Iraqi control.

Other military contributions: The Royal Navy and Air Force also undertook numerous missions to support Operation TELIC. Thereafter, the Royal Navy maintained a continuous presence in the North Arabian Gulf to defend Iraqi oil rigs and ensure the safe flow of supplies in and out of the region, as well as train Iraqi maritime security forces. The Royal Air Force flew thousands of sorties every year and also assisted in reconstruction of the Al Basrah area.

In April 2006, the United Kingdom established the Al Basrah Provincial Reconstruction Team with British, Danish, Australian, and Canadian participants. Its main activities centered on reestablishing local jurisdiction, civic institutions, and economic and social development. The team also attempted to build the capacity of the provincial government to manage public finances and the economy, including planning and development of budgets, with the overall aim of rebuilding local infrastructure and providing essential services such as sanitation, health care, and education.

Further Readings

Brown, Todd S. *Battleground Iraq: Journal of a Company Commander.* Washington, D.C.: Department of the Army, 2008.

Fontenot, Gregory; E. J. Degen; and David Tohn. *On Point: US Army in Operation Iraqi Freedom.* Fort Leavenworth, Kans.: Combat Studies Institute Press, 2004.

Hoffman, John T., gen. ed. *Tip of the Spear: U.S. Army Small-Unit Action in Iraq, 2004–2007.* Washington, D.C.: U.S. Army Center of Military History, 2009.

Joint Staff. Joint Publication 3–16, *Multinational Operations.* 7 March 2007.

Wright, Donald P., and Timothy R. Reese. *On Point II: Transition to the New Campaign: The United States Army in Operation Iraqi Freedom, May 2003–January 2005.* Fort Leavenworth, Kans.: Combat Studies Institute Press, 2008.

In addition to the volumes listed above, consult Congressional Research Service studies, Department of State Weekly Update briefings, Multi-National Force–Iraq briefings and press releases, and U.S. Central Command and Department of Defense briefings.

Abbreviations

AMTG	Al Muthanna Task Group
Br	British
CENTCOM	U.S. Central Command
CFLCC	Coalition Forces Land Component Command
CFSOCC	Combined Forces Special Operations Component Command
CJSOTF	Combined Joint Special Operations Task Force
CJSOTF-N	Combined Joint Special Operations Task Force–North
CJSOTF-W	Combined Joint Special Operations Task Force–West
CJTF-7	Combined Joint Task Force–7
CPA	Coalition Provisional Authority
EOD	explosive ordnance disposal
FOB	forward operating base
IED	improvised explosive device
IGC	Iraqi Governing Council
IRSAW	Iraq Reconstruction Support Airlift Wing
ISF	Iraqi Security Forces
IZ	International Zone
JGSDF	Japanese Ground Self-Defense Force
MNB PU	Multi-National Brigade Plus Ultra
MND-B	Multi-National Division–Baghdad
MND-C	Multi-National Division–Center
MND-CS	Multi-National Division–Center-South
MND-N	Multi-National Division–North
MND-NC	Multi-National Division–North Central
MND-NE	Multi-National Division–Northeast
MND-S	Multi-National Division–South
MND-SE	Multi-National Division–Southeast
MND-W	Multi-National Division–West
MNF-I	Multi-National Force–Iraq
MNF-NW	Multi-National Force–Northwest
MNF-W	Multi-National Force–West

MNSTC-I	Multi-National Security Transition Command–Iraq
NATO	North Atlantic Treaty Organization
NTM-I	North Atlantic Treaty Organization Training Mission–Iraq
SFIR	Stabilization Force Iraq
U.K.	United Kingdom
UN	United Nations
UNAMI	United Nations Assistance Mission for Iraq

MAP SYMBOLS

MILITARY UNITS

FUNCTION

Armor

Cavalry (Armored)

Infantry

Infantry (Air Assault)

Infantry (Airborne)

Infantry (Mechanized)

Marine Corps

SIZE SYMBOLS

Battalion or Cavalry Squadron   I I

Regiment or Group   I I I

Brigade   X

Division   X X

EXAMPLES

3d Infantry Division (Mechanized)

82d Airborne Division

101st Airborne Division (Air Assault)

173d Airborne Brigade

3d Armored Cavalry Regiment

3d Battalion, 75th Ranger Regiment

129

www.ingramcontent.com/pod-product-compliance
Lightning Source LLC
Chambersburg PA
CBHW070817100426
42742CB00012B/2385